SECOND EDITION

STATISTICAL THINKING
THROUGH
MEDIA EXAMPLES

BY ANTHONY DONOGHUE

Bassim Hamadeh, CEO and Publisher
John Remington, Executive Editor
Carrie Montoya, Manager, Revisions and Author Care
Kaela Martin, Project Editor
Alia Bales, Production Editor
Jess Estrella, Senior Graphic Designer
Alexa Lucido, Licensing Manager
Natalie Piccotti, Director of Marketing
Kassie Graves, Vice President of Editorial
Jamie Giganti, Director of Academic Publishing

Copyright © 2020 by Cognella, Inc. All rights reserved. No part of this publication may be reprinted, reproduced, transmitted, or utilized in any form or by any electronic, mechanical, or other means, now known or hereafter invented, including photocopying, microfilming, and recording, or in any information retrieval system without the written permission of Cognella, Inc. For inquiries regarding permissions, translations, foreign rights, audio rights, and any other forms of reproduction, please contact the Cognella Licensing Department at rights@cognella.com.

Trademark Notice: Product or corporate names may be trademarks or registered trademarks and are used only for identification and explanation without intent to infringe.

Cover images: Copyright © 2012 Depositphotos/ronleishman.
Copyright © 2010 Depositphotos/Arsgera.
Copyright © 2010 Depositphotos/jukai5.
Copyright © 2011 Depositphotos/lucadp.
Copyright © 2012 Depositphotos/leungchopan.
Copyright © 2012 Depositphotos/Andonde.
Copyright © 2012 Depositphotos/maxxyustas.
Copyright © 2012 Depositphotos/mishoo.
Copyright © 2013 Depositphotos/stori.
Copyright © 2013 Depositphotos/mmaxer.
Copyright © 2013 Depositphotos/minervastock.
Copyright © 2013 Depositphotos/Deklofenak.
Copyright © 2013 Depositphotos/scanrail.
Copyright © 2014 Depositphotos/bhofack.
Copyright © 2014 Depositphotos/Mactrunk.
Copyright © 2016 iStockphoto LP/Vertigo3d.

Printed in the United States of America.

Statistical thinking will one day be as necessary for efficient citizenship as the ability to read and write.

—H.G. Wells

CONTENTS

PREFACE — XI
ACKNOWLEDGMENTS — XV
INTRODUCTION — XVII

CHAPTER 1
STATISTICAL THINKING: WHY IS IT IMPORTANT? — 1

   1.1 Introduction — 1

   1.2 The MMR Vaccine-Autism Link — 3

   1.3 Samples and Populations — 4

   1.4 Selecting a Representative Sample — 6

   1.5 Observational Studies — 7

   1.6 Randomized Experiments — 9

   1.7 The Truth and Fake News — 11

   1.8 Conclusion — 12

   1.9 Real-World Exercises — 12

CHAPTER 2
ASSESSING THE QUALITY OF RESEARCH — 13

   2.1 Introduction — 13

   2.2 Alcohol Consumption and Cardiovascular Health — 15

   2.3 With Botox, Looking Good and Feeling Less — 22

2.4 The Fast Diet	26
2.5 American Football and Brain Injury	32
2.6 Conclusion	35
2.7 Real-World Exercises	35

CHAPTER 3
ASSESSING THE QUALITY OF POLLS AND SURVEYS   39

3.1 Introduction	39
3.2 Polling the BREXIT Referendum	41
3.3 Don't Ask, Don't Tell (DADT) Survey	46
3.4 Political Polarization and Media Habits	50
3.5 Asking Questions	52
3.6 Conclusion	53
3.7 Real-World Exercises	54

CHAPTER 4
VISUALIZING AND SUMMARIZING QUANTITATIVE DATA   57

4.1 Introduction	57
4.2 Histograms and the Normal Distribution	58
4.3 Normal Distribution: Measures Of Center and Spread	61
4.4 The Empirical Rule and the Standard Normal Distribution	64
4.5 Skewed Distributions: Measures of Center and Spread	67
4.6 Conclusion	72
4.7 Real-World Exercises	73

CHAPTER 5
MEASURING UNCERTAINTY WITH PROBABILITY — 75

    5.1 Introduction — 75

    5.2 Everyday Chance Events — 76

    5.3 What Is Chance? — 77

    5.4 Measuring THE CHANCES OF AN EVENT — 79

    5.5 Conditional Probability, BAYE'S RULE, and The Mammogram Controversy — 80

    5.6 A Tragic Statistical Error—The Case of Sally Clark — 84

    5.7 The Birthday Experiment — 85

    5.8 Conclusion — 86

    5.9 Real-World Exercises — 87

CHAPTER 6
VISUALIZING AND SUMMARIZING SAMPLE STATISTICS — 89

    6.1 Introduction — 89

    6.2 The Distribution of Sample Statistics — 90

    6.3 Measuring Variation in Sample Statistics — 94

    6.4 The Sample Size Condition — 106

    6.5 Predicting Election Results — 107

    6.6 Conclusion — 110

    6.7 Real-World Exercises — 110

CHAPTER 7
MAKING STATISTICAL DECISIONS WITH CONFIDENCE INTERVALS — 113

7.1 Introduction — 113

7.2 Confidence Interval for the Population Parameter — 114

7.3 Confidence Levels — 118

7.4 Constructing Confidence Intervals — 119

7.5 Conclusion — 132

7.6 Real-World Exercises — 132

CHAPTER 8
MAKING STATISTICAL DECISIONS WITH HYPOTHESIS TESTING — 135

8.1 Introduction — 135

8.2 Null and Alternative HypothesEs — 136

8.3 Population Proportion — 138

8.4 Population Mean — 141

8.5 What Does the P-value Really Mean? — 145

8.6 One-Sided versus Two-Sided Alternatives — 150

8.7 Population Mean Difference — 153

8.8 Analysis of Variance (ANOVA) — Comparing Several Means — 158

8.9 Population Proportion Difference — 161

8.10 Analysis of 2 x 2 Tables — Chi-Square Test — 169

8.11 Types of Error and The Power of the Test — 174

8.12 Conclusion — 182

8.13 Real-World Exercises — 183

CHAPTER 9
BUILDING ON FOUNDATIONS: LINEAR REGRESSION — 187

    9.1 Introduction — 187

    9.2 Visualizing Relationships Between Quantitative Variables — 188

    9.3 Correlation versus Causation — 190

    9.4 Simple Linear Regression — 196

    9.5 Confidence Intervals And Hypothesis Testing For The Population Slope — 205

    9.6 Multiple Linear Regression — 214

    9.7 Conclusion — 219

    9.8 Real-World Exercises — 219

CHAPTER 10
INTEGRITY IN RESEARCH — 223

    10.1 Introduction — 223

    10.2 Data Integrity — 224

    10.3 Replication Crisis in Psychology — 226

    10.4 The Low Power of Power Posing — 231

    10.5 Vioxx, Heart Attacks, and the Opioid Crisis — 235

    10.6 Conclusion — 241

APPENDIX — 243

    General Notation — 243

    Chapter 4 — 244

    Chapter 6 — 244

    Chapter 7 — 250

Chapter 8   253

Chapter 9   254

INDEX   **257**

PREFACE

When I sat down to write this book, it was with two purposes in mind. First, I wanted to write a book introducing the foundations in statistical thinking to a broader audience. Second, I wanted a support textbook for use in my introductory statistics courses. I needed a textbook that explained the conceptual meaning behind the mathematical calculations in a clear, concise, and engaging way. In a short one-semester course, there is simply not enough time in the classroom to do this. I needed a textbook that would support my classroom efforts by rigorously explaining the statistical thinking behind the calculations.

Statistics, the discipline, is not a branch of mathematics. It is the application of mathematics to data in order to turn data into meaningful information about how the world works. However, many students leave their first statistics course not realizing this fact. They learn how to make calculations like the p-value and use it to declare statistical significance when it is less than 0.05, yet they often leave the course with little understanding of what it all means. They may think they have proven something when all they have done is made a decision based in chance. In this day and age, computer software can calculate the statistics for us. It is our job to understand what the statistics really mean. It is our job to learn how to think statistically.

I fell in love with statistics and statistical thinking in my first days in college. My first statistics class was with Professor David Williams at the Statistics Department in University College Dublin in Dublin, Ireland. He told us that statistics calculated from a random sample of data were used to estimate some unknown truth in a larger population of interest. How close our estimate is to the truth (on average) will be the same no matter what the population size: 100,000, 1 million, 1 billion, or even 1 trillion. How close our estimate is to the truth is driven by the size of the random sample and not by the size of the population. I found this very intriguing. Soon after, I learned that the simple process of random assignment of patients to treatments enables researchers to make causal conclusions about which treatments are effective and which treatments are not. Without random assignment, there would be no way of knowing. Again, the power of a random process to get at the truth. I was hooked!

Through studying and teaching the discipline of statistics, I have learned how to think critically about the information I take in through the media. I have learned how to go beyond the news headlines to the source of the research, poll, or survey to critique the quality of the statistical evidence for myself. I have learned how to question the quality of study design and the data the research is based upon. I have learned that most data are observational in nature where cause and effect conclusions can't be made no matter what the news headlines might say. I have learned that correlation does not mean causation in observational studies due to confounding factors. I have learned how to understand and reason with the complex nature of variation in measurements and the relationships between different types of measurements we observe in the world around us every day. I have learned that Statistics, the discipline, is the engine that drives decision-making in any science that involves the analysis of data. I have learned that for the challenging questions that science asks, truth is a pursuit, not a destination. I have learned that the distance between the truth and our estimate of truth, the statistic, can be minimized by collecting quality data. I have learned how to reason with the uncertainty regarding what the truth may be using probability and statistics. I have learned that the path to truth is steep with many potential pitfalls along the way. I have learned to applaud those researchers who are willing to do the hard and honest work of pursuing what is true and not what they want to be true. I have learned how the statistical method can be manipulated, misused, or misunderstood to the advantage of the researcher. I have gained the insight that there is no path to truth without integrity. I have learned a dedication to pursuing truth in science (and in one's life) has its own rewards that can't be measured or quantified. Ultimately, the discipline of Statistics has taught me a great deal about how to question, how to think, and how to live in a complex, chaotic, and uncertain world.

With this knowledge came a teacher's passion to share with others. In this book, I strive to explain the foundations in statistical thinking in a clear, concise, and engaging narrative. We begin with a classic example where, aided by the media, an unethical researcher cast doubt on what was considered an accepted truth: the safety of vaccines. This case, which helped give rise to the antivaccine movement, shows that a basic level of statistical thinking skills could have helped members of the media avoid disseminating information that was not only incorrect but actually harmful to the public.

Throughout this book the statistical concepts are explained through the context of media examples that are topical and related to our everyday lives. In my experience, this makes the concepts more relevant and easier to digest. Through media examples and case studies, I will lead you through the foundations in statistical thinking one step at a time. For many of you, the path through the first five chapters will be challenging but manageable. Upon completing these chapters, you will have learned how to think critically about the quality of data and study design, including polls and surveys; how to understand and reason with variation in data measurements; the basics rules of probability and how they relate to real world issues like the prevalence of false positives in mammogram testing. However, on our journey to thinking statistically, we've covered only half the distance. We climbed the foothills of statistical fundamentals but ahead still lies our summit, no less a goal than the search for truth through data and statistics.

In Chapters 6 through Chapter 8, we lay down the framework and methods used for how decisions are made using data and statistics. There is a beauty in this process but like many beautiful things, it is not acquired easily. Statistics is a rigorous craft. It requires you to reason through the steps in the process with patience and persistence. At times, the steps can be steep and the journey arduous. It is a repetitive process like learning chords on a guitar. However, that is how the concepts get engrained in your mind. I am especially careful when it comes to explaining the p-value and what it really means to declare statistical significance when it is less than 0.05. I will also discuss the movement within the statistical and scientific community towards abandoning this decision rule for declaring statistical significance altogether. Statistical analysis is the art of reasoning with uncertainty in order to make decisions regarding truth. However, our practical need to draw a definitive conclusion from an uncertain process can cause confusion. I will guide you through the entire process slowly and carefully, one step at a time.

In Chapter 9, we will build on your newly acquired foundations in statistical thinking by introducing a powerful and sophisticated technique for modeling the complex nature of relationships between different types of measurement data in the world around us. This knowledge is essential for those of you who want to go on to become a statistical analyst, researcher, or data scientist.

Our final chapter completes our journey to truth through data and statistics. Here, through your critical and statistical thinking skills, you will learn how to determine when methods of statistical analysis are misused, manipulated, or simply misunderstood. You will see that there is no path to truth without integrity by examining a case study where the lack of integrity on the part of individuals at a large pharmaceutical company led to tens of thousands of unnecessary deaths. As in the vaccine example of Chapter 1, if some members of the media with the understanding of statistics laid out in this book had questioned the results of the research, they might have prevented at least some of those deaths.

The need to question the world around us and to think for ourselves has never been more necessary. As Americans, we have the fundamental right to question our authority figures in order to get at the truth. It is essential for maintaining a healthy democracy. However, we need critical and statistical thinking skills to be able to ask the right questions in order to see through the noise and get at the truth. The truth will always be uncertain and that makes our quest all the more challenging. However, we can learn how to face, embrace and reason with that uncertainty in order to make good decisions. This book will train your mind in the sort of rigorous thinking required to do just that.

Never before have so many people had such a world of information at their fingertips. We can be mindless consumers and let other people tell us what to believe or we can learn how to separate good information from bad and understand how to use data to question the world around us. While it can be challenging to train your mind to move from the former to the latter, I believe you will find it well worth the effort. As a teacher, I really hope you make this journey with me and see the need for learning how to pursue truth through statistical thinking. If you are up for it, let's get to work!

ACKNOWLEDGMENTS

In the Statistics department at Columbia University, I would like to thank the following: David Madigan for the original opportunity to teach for the department, recognizing my suitability to teach statistical thinking to a broader audience, and for your comments on the preliminary edition; I want to thank Dood Kalicharan, Richard Davis and Tian Zheng for your constant and continued support; I want to thank Andrew Gelman for your comments on the second edition. I would also like to thank Sander Greenland at UCLA and Valentin Amrhein at the University of Basel for your comments on the second edition.

I want to thank my wife, Susan Donoghue, for reading and editing early versions of chapters, helping me set the tone of the narrative, and for giving me the necessary space and support to write this book. I would like to express my gratitude to my good friend Dan Gronell, whose comments and edits on the preliminary edition were so valuable and insightful. I want to thank my good friend Colin Merry, for your encouragement and support over the years, and for always believing that this book was a great idea.

I want thank the students who helped bring this book to fruition; I want to thank Justinas Grigaitus for your excellent feedback on the more challenging chapters. I want to thank Martin Deeb for your insightful comments. I want to thank Samuel Oh for your last minute contribution that was so valuable. I want to thank Shandu Mulandzi for your excellent work on the lab component that accompanies this book. I want to thank Loren Crabbe for your excellent work on the process maps and select graphics in the slides that accompanies this book. I want to thank every student I have taught over the years that took on the challenges I put in front of them. Every insight and explanation contained in this book grew out of my many years down in the trenches with you.

Finally, I want to thank everyone at Cognella Publishing involved in the production of this book. Thank you for giving me the creative freedom to find my voice, and the necessary support and time to bring forth the book I envisioned.

INTRODUCTION

To know God's thoughts one must study Statistics, for these are a measure of His purpose.

—*Florence Nightingale*

On May 12, 1820, Florence Nightingale was born into a very wealthy British family. As a Unitarian, she was brought up to believe her faith was a call to action to help those less fortunate than herself. As a woman, she was expected to live a life of leisure: dancing, vacations, and attending society functions. As a strong minded and empathetic individual, these two conflicting points of view were destined to cause a stir in her.

At the age of seventeen, she felt a calling to pursue a "great quest." However, she did not know what exactly she was supposed to do. She continued to live the life she was expected to live until her mid-twenties, when she came to the realization she wanted to become a nurse. At this time nursing was considered a very lowly profession with a reputation for drunkenness and prostitution. Hospitals were filthy places where few people went and far fewer returned.

Nightingale was ready to pursue her chosen path with a passion. However, as expected, her family was very resistant. She never gave up on the idea, and eventually, at the age of thirty-one, her family allowed her to get the training she needed at the Institution of Deaconess in Kaiserswerth, Germany.

Nightingale's reputation as an outstanding nurse and administrator grew very quickly. In March 1854, England went to war with Russia in what became known as the Crimean War. Reports coming back from the front shocked the British public, with thousands of men dying from diseases like cholera and typhus. Into the mix was thrown thirty-four-year-old Florence Nightingale. She would lead a team of almost forty nurses to a hospital in Scutari, Turkey, as the superintendent of the female nursing establishment of the English General Hospitals in Turkey.

The conditions she and her team of nurses found at the hospital were appalling: filth everywhere, with the soldiers receiving terrible care and very poor diets. Nightingale wasted no time turning things around. She had the common sense to know the importance of hygiene in hospitals several decades before the science of germ theory. Her great empathy for the soldiers—by day implementing changes to improve their living conditions, by night visiting and comforting them—led her to become affectionately known as "The Lady of the Lamp."

It was during this time that Nightingale discovered her love and appreciation for data and statistics. She kept detailed records on how the soldiers died and what they died from: preventable diseases, the result of wounds, and other causes. In total, 20,400 soldiers died in the Crimean War, with 16,000 dying from preventable illness and poor care. The death toll would have been much greater were it not for the great work of Nightingale and her team of nurses.

On her return to England, Nightingale worked with one of the few statisticians at the time, William Farr, to present her findings to the British government to push for reform. By analyzing the data Nightingale collected, they were able to show (for example) statistical evidence that soldiers who were huddled together in small areas were more likely to die than soldiers that were spread far apart.

Her passion for data and statistics continued to grow. Nightingale had an astute understanding of the power of graphical presentations to communicate the reality she saw on the ground in Turkey. She invented a graphic she called the "coxcomb," a very creative type of pie chart. She used her graphical presentations of how the soldiers died to persuade the government to implement the changes in hospital administration necessary to save lives: good hygiene, ventilation, and nutrition. With perseverance and by overcoming many obstacles, she eventually got to see the changes she wanted implemented. The rest, as they say, is history.

Statistics are a measure or estimate of truth in the world around us. If data is collected and analyzed properly, the resulting statistics can be very good estimators of truth. Nightingale was one of the first individuals to understand the power of data collection and analysis as a means of pursuing truth and bringing about real change. Her calling to do good in the world was greatly aided by her understanding and appreciation of the power of statistics and statistical thinking.

Since the days of Nightingale, statistics has provided us the tools to reason with the complex, chaotic and uncertain nature of the world around us. The use of statistics has led to advances in many different fields—agriculture, psychology, medicine, finance, business, astronomy, economics, sociology and political science. In medicine, statistics has enabled us to make cause-and-effect conclusions regarding what treatments work and don't work, advancing the well-being of individuals throughout the twentieth century and into the twenty-first century.

Statistical thinking is a common-sense, observational, and fact-based way of thinking about the world that will continue to bring understanding and insight into how the world works. Statistics are estimators of truth, but good statistics are more than good enough to help us understand how to live healthier, more enlightened, and informed lives.

CHAPTER 1

STATISTICAL THINKING: WHY IS IT IMPORTANT?

The fabulous statistics continued to pour out of the telescreen. As compared with last year there was more food, more clothes, more houses, more furniture, more cooking pots, more fuel, more ships, more helicopters, more books, more babies—more of everything except disease, crime, and insanity. …

— *George Orwell*, Nineteen Eighty-Four

1.1 INTRODUCTION

In his book *Nineteen Eighty-Four*, George Orwell imagined a stark future where the vast majority of the population are controlled by a privileged few belonging to the Inner Party. The leader of the party, known as Big Brother, is omnipresent, and anyone showing outward signs of individual thought or expression is persecuted.

The protagonist, Winston, works for the Ministry of Truth. His job is to rewrite history to conform to what the state wants its citizens to believe. He starts to get frustrated with his position and the manipulation of truth by the ministry.

George Orwell conceived of the novel during World War II, a time of great uncertainty in world history, and completed the novel during the early years of the Cold War. It was first published in 1949. The novel chillingly portrays how the thoughts and emotions of individuals can be manipulated by information they consume every day, when presented by authority figures.

KEY TERMS

Population: A large group of individuals with a particular characteristic of interest

Sample: A subgroup of the population actually studied

Population Parameter: An unknown quantity that, for a given population, is fixed and is used as a value to describe a particular characteristic of the population (e.g., average height)

Sample Statistic: An estimate of the population parameter taken from a sample of measurement values (e.g., average height)

Variable: Characteristic recorded about each individual (e.g., height, age, race)

Representative Sample: A sample of individuals from a population that is representative or matches the population (in terms of demographics and other characteristics) as much as possible

(Simple) Random Sample: A sample of individuals that are representative of the population of individuals because each individual in the population is equally likely to be selected to be in the sample

Convenience Sample: A sample of individuals who are easy to reach

The sample is highly unlikely to be a representative sample

Average: A numeric value representing the central or typical value in a set of measurements

Observational Study: Study where individuals are not assigned to treatment groups; they are simply observed and individual measurements are collected

Confounding Factor (or Variable): A factor (besides the treatments) that differs between groups under study that affects the variable being measured.

Randomized Experiment: Research where individuals are randomly assigned to different treatments. This process ensures that all possible (confounding) factors are averaged out. As a result, any differences in sample statistics (between treatments) can be said to be due to the different treatments

Statistically Significant: A declaration that is made when the p-value is considered small (usually less than 0.05)

Sample Effect Size: When comparing treatment groups, it is the size of the difference of sample estimates between treatment groups

Placebo: An inactive (dummy or fake) treatment used as a control in an experiment

Sampling Variation: The variation in sample statistics from sample to sample

Statistical Decision: A decision based in chance regarding the value of a population parameter

We rely on authority figures and other individuals for information on what to think, believe, and do. These include parents, high-school teachers, professors, priests, rabbis, imams, journalists, advertisers, and politicians. We build our belief system about how the world works based on what our most influential authority figures tell us. Once we establish a belief system about how the world works that we are comfortable with, we have a tendency to not be very open to other points of view.

However, relying on such individuals for truth can be dangerous. They may not have your best interests at heart or be very knowledgeable in the subject they are talking about. When it comes to the news media, according to the journalist James Hamblin in the *Atlantic*:

1984 GEORGE ORWELL MOVIE TRAILER

Web Link: https://www.youtube.com/watch?v=Z4rBDUJTnNU

Search Term: 1984 George Orwell Movie Trailer

WHAT DO YOU MEAN BY THE MEDIA?

Web Link: https://www.theatlantic.com/technology/archive/2017/01/all-possible-realities-are-playing-out-across-infinite-universes/514130/

Search Term: What Do You Mean by the Media?

What Do You Mean by the Media?

The goal in journalism is to be the best at identifying and conveying said truth. The entire concept of the profession is antithetical to lying. So it's difficult to imagine objecting to the idea of journalism, in principle: to have people whose job is to act as dispassionate arbiters who discern truth. People who are fair, who are trustworthy, who do not slander, who are not beholden to any particular interest but seek transparency, to highlight injustice, and to hold people in power accountable.

STATISTICAL THINKING: WHY IS IT IMPORTANT? | 3

This is a worthy goal that every journalist should aspire to. The problem is that journalists are in the business of selling news and sensational stories that may not be based in truth or good science sell news.

In this chapter, we will discuss a classic case of where the media failed to investigate the validity of a piece of research before reporting on it. We will begin to understand why it is so important to question the information presented to us in the media every single day. In the process, we will answer the following questions:

- What do we mean by a sample of individuals versus a population of individuals?
- What do we mean by a sample statistic versus a population parameter?
- Why is sample size and the quality of the sample selected important?
- How can researchers select a representative sample of individuals from a population?
- How do researchers determine cause-and-effect relationships in a population of individuals?
- What do we mean by confounding factors or variables?
- What is the difference between an observational study and a randomized experiment?

1.2 THE MMR VACCINE-AUTISM LINK

In 1998, Andrew Wakefield, a well-respected doctor at the time, presented the results of his research in the *Lancet*, one of the world's leading peer-reviewed medical journals, suggesting a link between the measles, mumps, and rubella (MMR) vaccine and autism.

ABC did a report in 2000 on Wakefield's findings. On the surface, the report seems fair and balanced, interviewing experts on both sides of the debate. However, a simple review by the reporter of the actual journal article where Wakefield presented his findings would have ended the debate at that time. The title of the journal article presented in the *Lancet* was "Ileal-Lymphoid-Nodular Hyperplasia, Non-specific Colitis, and Pervasive Developmental Disorder in Children."

First, Wakefield strongly suggests a link between the MMR vaccine but does not make a causal conclusion. Association does not mean causation. In the journal article he states:

> *Rubella virus is associated with autism and the combined measles, mumps, and rubella vaccine (rather than monovalent measles vaccine) has also been implicated. Fudenberg noted that for 15 of 20 autistic children, the first symptoms developed within a week of vaccination.*

ABC REPORT 2000: MMR-AUTISM LINK

Web Link: http://www.youtube.com/watch?v=IhL6Sl2zhYo

Search Term: ABC News Autism 2000

Second, the research was based on only twelve children vaccinated with MMR:

> *12 children (mean age 6 years [range 3–10], 11 boys) were referred to a paediatric gastroenterology unit with a history of normal development followed by loss of acquired skills, including language, together with diarrhea and abdominal pain. Children underwent gastroenterological, neurological, and developmental assessment and review of developmental records.*

ILEAL-LYMPHOID-NODULAR HYPERPLASIA, NON-SPECIFIC COLITIS, AND PERVASIVE DEVELOPMENTAL DISORDER IN CHILDREN

Web Link: http://www.thelancet.com/journals/lancet/article/PIIS0140-6736(97)11096-0/abstract

Search Term: Retracted Andrew Wakefield

Wakefield found that nine of the children went on to develop autism soon after receiving the vaccination. Though his conclusions were speculative and based on only twelve children, this did not stop the media from widely reporting that Wakefield had found a causal link between the vaccine and autism.

Parents depend on the media to act as gatekeepers when it comes to these sorts of controversial claims made by researchers. They lead busy lives so they need to trust that the media is critiquing the quality of research before presenting the findings to the public. When it comes to the health and well-being of their children, a proper critique of the findings is more helpful than a sensational story.

If the news media had taken the time to read in the journal article that Wakefield's conclusions were speculative and based on only twelve children, the reporting of the research would (or should) never have happened. Strong claims require strong evidence. A claim that the MMR vaccine is associated with autism based on just twelve children is not strong evidence.

Small sample sizes can be useful if natural variation in measurement values is relatively small and the sample is representative of the population it was selected from. In Chapter 2, we will discuss the findings of a well-conducted study looking at relationship between football playing and brain injury that was based on a small sample size. However, as we will learn, Wakefield's sample of twelve children was far from representative of all children in the United Kingdom (UK) who received the MMR vaccine, and therefore poor quality sample data.

1.3 SAMPLES AND POPULATIONS

Statistics and statistical thinking are about trying to gain an understanding of the characteristics of **populations**. In the Wakefield study, the population of interest was all children in the UK who had received the MMR vaccine. Wakefield wanted to estimate the percentage of these children who went on to develop autism. In order to estimate this percentage, a **sample** of children who received the MMR vaccine was selected. The percentage of children in this sample who developed autism is called a **sample statistic**. It is an estimate of what is known as the **population parameter** of interest. In this case, the population parameter of interest is the true percentage of children in the population of MMR-vaccinated children who went on to develop autism.

Researchers are interested in looking for relationships between variables in the population of interest. A **variable** is simply a characteristic about an individual that we measure. In this study, the characteristics were MMR vaccine (yes or no) and autism (yes or no). Whether or not a child received the MMR vaccine is called the explanatory variable. It is used to try and explain the outcome or response variable, whether or not the child went on to develop autism. Wakefield was interested in the relationship between autism and exposure to the MMR vaccine in the population of children in the United Kingdom. Are children who receive the vaccine more likely to get autism?

Wakefield found that 75% or nine of the twelve MMR-vaccinated children he sampled went on to develop autism. If this percentage (the sample statistic) was anywhere close to the true percentage in the population (the population parameter), then there would have been a noticeable increase in the incidence of autism after the vaccine was introduced. Wakefield discusses this fact near the end of the journal article.

> *If there is a causal link between measles, mumps, and rubella vaccine and this syndrome, a rising incidence might be anticipated after the introduction of this vaccine in the UK in 1988. Published evidence is inadequate to show whether there is a change in incidence or a link with measles, mumps, and rubella vaccine.*

The fact there was little evidence of an increase in the overall percentage of children with autism since the vaccine was introduced is another reason why the news media should not have reported on this research in the way that it did.

What role does sample size play in how close our sample statistic is to the population parameter of interest?

We will begin to think about this question using the example of a coin toss experiment. Say we want to test whether a coin is fair. If the coin is fair, then we expect it to land heads 50% of the time. We can think of this percentage as our population parameter of interest. We can think of the percentage of heads in our sample of coin tosses as our sample statistic.

Say you toss a coin 10 times. If the coin is fair, we expect to get 5 (50%) heads, but if we got 3 (30%) heads or 7 (70%) heads, we would not be that surprised. We would not be ready to say that the coin is not fair. However, if we toss the coin 100 times and get 30 (30%) heads or 70 (70%) heads, we might start to feel like the coin is not fair.

A small sample size will (more often than not) result in a sample statistic that is far from our parameter of interest. In other words, statistics are highly variable estimates of population parameters when the sample size is small. However, as the sample size increases, we expect our sample statistic to converge toward the population parameter of interest. In other words, the larger our sample size, the closer we expect the sample statistic to be to the truth.

Andrew Wakefield was interested in determining the percentage of children in the population who received the MMR vaccine who went on to develop autism. A sample size of twelve children, even if it were a properly selected representative sample of children from the population, is unlikely to give a reliable estimate of that percentage.

1.4 SELECTING A REPRESENTATIVE SAMPLE

When selecting a sample from some population, we need to select a **representative sample**, a sample that gives us a fair and unbiased estimate of the population parameter of interest, for example the **average** height of men in the United States. A **(simple) random sample** is expected to be a representative sample. A (simple) random sample is one in which every individual in the population has an equal chance of being selected. Obtaining a proper random sample of individuals is often easier said than done.

For example, let's say you want to determine the average height of the population of male students at your college. At mid-day, you decide to stand in the middle of your campus, where students are heading to and from their classes, asking male students their height as they pass by. You believe that all male students pass by (where you are standing) at some point during any given day. Therefore, you feel that your sample should be random and therefore a representative sample of all male students.

However, what if on that particular day and time the male basketball team just got back from a game and were passing by. Including these men in your sample would result in an overrepresentation of tall men in your sample. In other words, there would be a higher proportion of tall men in your sample than in the population. The sample will result in an estimated average height well above the population average.

By selecting a random sample, we are attempting to obtain a sample where every individual is equally likely to be selected, resulting in a representative sample of individual heights. This representative sample of individual heights will ensure a sample average height closer to the population average height than if we were to include the basketball team in the calculation. With a properly selected random sample, the larger the sample size, the closer we expect our sample average will be to the population average.

This is the power of a random sample, and it is one of the most important concepts at the heart of statistical thinking. In addition, the size of the population does not matter. It can be one hundred thousand, a million, a billion or even a trillion. How close our sample statistic is to the population parameter of interest is driven by the size of a properly selected random sample.

In the MMR-autism study, the sample size was small and far from representative. In January 2011, the investigate journalist Brian Deer published a paper in the *British Medical Journal* titled "How the Case Against the MMR Vaccine Was Fixed," presenting the results of his investigation into Wakefield.

Brian Deer did an exhaustive job of investigating the truth in this case. A summary of his findings can be found on his website. He found that two years before Wakefield published his findings (and had selected the twelve children for his study), he was hired by a lawyer named Richard Barr to attack MMR. Brian Deer also discusses how Wakefield selected his sample of twelve children.

The type of sample that Wakefield selected is known as a **convenience sample**. In his case, the sample was conveniently chosen in a way to ensure he showed evidence of an association between autism and the MMR vaccine. We will learn that random samples can be difficult to obtain and researchers will often rely on convenience samples. How useful a sample is in determining population characteristics depends on how far from representative of the population it is.

1.5 OBSERVATIONAL STUDIES

The type of study that Andrew Wakefield conducted is known as an **observational study**. In a properly conducted observational study, there is no manipulation of the subjects involved. The subjects are simply observed in their environments and measurements are taken. We are looking for an association between exposure to a factor and the outcome. In the Wakefield study, he was looking for an association between exposure to the MMR vaccine and whether or not a child went on to develop autism.

We cannot make causal conclusions when analyzing data from an observational study due to what are known as **confounding factors (or variables)**. Every child is born to different parents and into various types of environments. There are many factors (genetic, environmental, etc.) that may differ substantially between children who develop autism and those who do not.

The real challenging question is how do researchers figure out what these factors are? What are the factors that are causing autism? The Autism Society of America provides information on their website regarding the causes of autism. At this stage, researchers have concluded that the causes of autism are most likely genetic in nature and not environmental.

HOW THE CASE AGAINST THE MMR VACCINE WAS FIXED

Web Link: http://www.bmj.com/content/342/bmj.c5347

Search Term: How the Case Against the MMR Vaccine Was Fixed

EXPOSED: ANDREW WAKEFIELD AND THE MMR-AUTISM FRAUD

As with the researcher, so too with his subjects. They also were not what they appeared to be. In the *Lancet*, the 12 children (11 boys and one girl) had been held out as merely a routine series of kids with developmental disorders and digestive symptoms, needing care from the London hospital. That so many of their parents blamed problems on one common vaccine, understandably, caused public concern. But Deer discovered that nearly all the children (aged between 2½ and 9½) had been pre-selected through MMR campaign groups, and that, at the time of their admission, most of their parents were clients and contacts of the lawyer, Barr.

Web Link: http://briandeer.com/mmr/lancet-summary.htm

Search Term: Exposed: Andrew Wakefield and the MMR-Autism Fraud

> **CAUSES OF AUTISM**
>
> There is no known single cause for autism spectrum disorder, but it is generally accepted that it is caused by abnormalities in brain structure or function. Brain scans show differences in the shape and structure of the brain in children with autism compared to in neurotypical children. Researchers do not know the exact cause of autism but are investigating a number of theories, including the links among heredity, genetics and medical problems.
>
> In many families, there appears to be a pattern of autism or related disabilities, further supporting the theory that the disorder has a genetic basis. While no one gene has been identified as causing autism, researchers are searching for irregular segments of genetic code that children with autism may have inherited. It also appears that some children are born with a susceptibility to autism, but researchers have not yet identified a single "trigger" that causes autism to develop.
>
>
>
> Web Link: http://www.autism-society.org/what-is/causes/
>
> Search Term: Causes, Autism Society of America

It is the variability in factors (genetic, heredity, environmental and numerous others) from person to person that makes it very difficult to make a cause-and-effect conclusion from studies using observational data. As a result, there is a level of uncertainty inherent in the results of observational studies that is very difficult for researchers (and the media) to communicate to the general public and for the general public to accept and understand.

Another good example is the effect of alcohol consumption on our health and well-being. In 2018, two very large observational studies that received a lot of media attention looked at the relationship between increasing levels of alcohol consumption and health outcomes: heart disease and life expectancy. From reading the news headlines, one would conclude that the optimal amount of alcohol consumption is none at all. We will question the results of one of these studies in depth in our next chapter. For now, we will simply state that when studying alcohol consumption using observational data, it is impossible to get the effect of every possible confounding factor out of the way so as to clearly see the health effects of increasing levels of alcohol use. There are many factors in the lifestyle of the individual (level of exercise and diet, for example) that may be related to the explanatory variable (level of alcohol use) and affect the response variable (like heart disease). It is impossible to measure and control for all these factors from individual to individual in an observational study.

A report by the *New York Times* titled "Vaccines: An Unhealthy Skepticism/Measles Virus Outbreak 2015/Retro Report" sums up very well what we have been discussing. It talks about the MMR-autism controversy, including a discussion about the difficulty that scientists and journalists face in communicating the uncertainty inherent in the results of scientific research.

In June 2017, *Last Week Tonight* with John Oliver did an educational and amusing piece on the state of vaccines. He opens with the fact that vaccines have saved millions of lives, looking back at a time when people lined up for the polio vaccine like they do for iPhones today. However, he goes on to point out how the skeptics are continuing to gain more of a voice in the debate, with increasing numbers of parents deciding not to vaccinate their children.

> **ARTICLE: MEASLES CASES AND OUTBREAKS**
>
>
>
> Web Link: https://www.cdc.gov/measles/cases-outbreaks.html
>
> Search Term: Measles Cases and Outbreaks

According to the Center for Disease Control (CDC), from January to May 10th 2019, 839 cases of measles were confirmed across 20 states. This is the largest number of measles cases in any given year since 2000, the year when measles was declared eliminated and the media started to sensationalize the results of Andrew Wakefield's research.

1.6 RANDOMIZED EXPERIMENTS

From the discussion so far, you may be asking yourself some or all of the following questions:

- If it is so difficult to make cause-and-effect conclusions, how do researchers decide what medications are effective for treating a particular disease or condition?
- How do researchers decide whether a treatment say for lowering cholesterol is more effective than simply giving the patients a **placebo** (dummy or fake) treatment?
- How do researchers know whether it was the treatment that lowered the patient's cholesterol level and not some other factor (or factors) that affects their cholesterol level?

The type of study that enables us to find answers to these questions is known as a **randomized experiment**, often called the gold standard in research. A properly conducted well-designed randomized experiment has the power to allow us to make causal conclusions about the effectiveness of a drug. If the **sample effect size** is found to be **statistically significant**, then we can make a causal conclusion. The design of the study has the power to eliminate or average out other factors (besides the treatment) that may affect the patient's response or measurement value.

For example, the Mayo Clinic website lists five ways to improve cholesterol levels. A participant of a study receiving a treatment to reduce cholesterol levels may decide to eat more healthily, exercise, and/or stop smoking. How can the researcher know whether it was these activities or the treatment that lowered the patient's cholesterol level? Another problem could be that there are more (or fewer) patients who eat healthily or exercise on the drug than on the placebo or vice versa.

The researcher can solve these problems by designing their study as a randomized experiment. The researcher can randomly assign patients to treatments—say active treatment

VACCINES: AN UNHEALTHY SKEPTICISM | MEASLES VIRUS OUTBREAK 2015 | RETRO REPORT

Web Link: https://www.youtube.com/watch?v=fMsa7o48XBE

Search Term: Vaccines: An Unhealthy Skepticism

VACCINES: LAST WEEK TONIGHT WITH JOHN OLIVER (HBO)

Web Link: https://www.youtube.com/watch?v=7VG_s2PCH_c

Search Term: John Oliver Vaccines

TOP 5 LIFESTYLE CHANGES FOR REDUCING CHOLESTEROL LEVELS

1. To eat what are known as "heart-healthy" foods, such as berries and nuts.
2. To exercise daily or almost daily and in general to increase your overall amount of physical activity.
3. To stop smoking.
4. To lose weight.
5. Limit the amount of alcoholic beverages you drink.

Web Link: http://www.mayoclinic.org/reduce-cholesterol/art-20045935

Search Term: Mayo Clinic Cholesterol Lifestyle

and placebo. Random assignment has the power to remove (or average out) the effects of any other factors on cholesterol level. For example, let's say we had a sample of one hundred subjects with high cholesterol. We want to test our new treatment for lowering cholesterol by giving half the subjects the treatment and the other half a placebo. To remove the effects of confounding factors, the subjects are randomly assigned to the treatment and placebo. This can be done using a random number generator on your computer. The key to proper random assignment is that each subject is equally likely to be assigned to either treatment.

Random assignment should ensure that all other confounding factors, which may affect a subject's cholesterol level, will be averaged out. This means that subjects assigned to either treatment or placebo should be approximately the same average age, weight, height, and so on. There should be approximately the same proportion of subjects who smoke, drink alcohol, and exercise on the treatment and the placebo.

After random assignment is completed, the only factor that should differ between the two groups is the treatment that they are on. If we find that the average cholesterol level is statistically significantly lower for patients on our treatment compared to patients on placebo, then we can say that the difference is most likely due to the treatment.

Random assignment is a very powerful technique. However, it does not eliminate all the uncertainty in our decision making. When we say that the results of our randomized experiment are statistically significant, we are stating that given the sample effect size (from the sample data we collected), there is a very small chance that our treatment would not be effective (on average) if the treatment were given to everyone in the population. Therefore, based on the statistical evidence found in our sample data, we decide that the treatment is effective. However, it does not mean we definitely made the right decision.

Why are we still uncertain as to whether we made the right decision about the effectiveness of our drug? The uncertainty remains because we only used a sample of one hundred subjects with high cholesterol out of a much larger population of people with high cholesterol. The sample only gives us an estimate of the average effect of the drug on the entire population, and different samples will result in different sample statistics, all estimates of the same population effect.

This concept is known as **sampling variation**. Sample statistics vary from sample to sample and all are estimates of the unknown population parameter of interest. Later in the book, we will learn how to reason with sampling variation in order to make a **statistical decision** with regard to the possible value of the population parameter of interest.

Every piece of research starts with a challenging question about the world around us—about truth in populations. The pursuit of an answer will lead us to the insight that we can't know what the truth is. However, through the use of statistics and statistical thinking we can measure, minimize, reason with, and communicate the level of uncertainty we have with regard to what the truth may be. By avoiding error and bias in our measurements, we can minimize the distance between the statistic, our estimate of truth, and the truth itself.

1.7 THE TRUTH AND FAKE NEWS

The rapid growth in ways people digest their news has made it much more difficult to separate fact based news from fake news. With so many news sources, it can be hard to know what to believe. As stated earlier in the chapter, we have a tendency to believe what we want to believe, a tendency that can be easily manipulated.

Having a healthy skepticism about what we read in the media is a good thing. There is an old adage that says "Paper won't refuse ink," meaning we should always be skeptical about what we read in the media. However, we also need to be willing to think critically about the news we read or listen to every day so we can learn how to separate fact based news from fake news.

This requires the use and development of our critical thinking skills. Critical thinking is about making a well-reasoned judgement about the information we take in. In the past, this was easier to do. Up until the 1980s, most people turned on the six o'clock news (mainly ABC, CBS, or NBC) to listen to news anchors like Walter Cronkite on CBS News. The news anchors could be trusted to present the facts. Afterward, people discussed the facts and drew their own conclusions. Knowing the facts did not eliminate our anxiety regarding the state of the world. However, good or bad, at least we could trust the information we were presented with.

In the Information Age, this is no longer the case. With so many news sources to choose from, we have to question their credibility. We want to trust in our favorite news source like people used to trust Walter Cronkite (affectionately known at the time as "Uncle Walter"). Unfortunately, not everyone working in the news media today have his level of integrity.

Therefore, we need to question the quality of our sources of news and information for ourselves. As the opinion piece in the *New York Times* titled "Dr. Google is a Liar" points out, critical thinking has become necessary not only for the health of our democracy but also for the health of our individual selves. There is a lot of misinformation on the Internet about the risks of different medications (including vaccines) playing on people's fears as Andrew Wakefield once did and continues to do so.

The opinion piece also points out that big tech and journalists need to do a better job of presenting more factual information to the public, and researchers need to communicate key concepts (like the difference between an observational study and a randomized experiment) in a more engaging way.

In this book, we will learn how to pursue truth by developing our critical and statistical thinking skills. We will develop our critical thinking skills by going beyond the news headlines to the source of the research to learn how to tell good research methodology from bad. We will learn the foundations of statistical thinking used for pursuing truth with data and statistics. In the process, we will learn how to think for ourselves and how to pursue truth in the complex and chaotic world of mass information.

NEWS ARTICLE: DR. GOOGLE IS A LIAR

Web Link: https://www.nytimes.com/2018/12/16/opinion/statin-side-effects-cancer.html

Search Term: Dr. Google is a Liar

1.8 CONCLUSION

In this chapter, we learned why it is so important to question the information we take through the news media every day. Journalists may not have the knowledge, understanding, or inclination to question the results of research put forth by scientists. We went beyond the news headlines to the source of the research to question the vaildity of the research for ourselves.

In the process, we learned some basic but very important concepts in statistical thinking. In the next chapter, we will continue to develop our critical and statistical thinking skills by critiquing journal articles related to other examples of research presented in the media.

1.9 REAL-WORLD EXERCISES

1. Find a story in the media that discusses the results of a piece of research. Determine whether the researchers completed a randomized experiment or an observational study. Based on the concepts we have discussed in this chapter, decide how seriously you should take the results of the research.

2. Select two media outlets, one that is considered to be part of the "liberal media," the other part of the "conservative media." Find a news story in each outlet on a topic that is considered controversial, like global warming. Discuss any differences you observe in how the facts of the story were presented by the two media outlets.

3. Go to the website Our World in Data at www.ourworldindata.org. Select a topic of interest to you. Select two of the data visualizations presented under your chosen topic. Write a brief summary of what you see in the data visualizations.

4. Referring to Q3, download an Excel data file related to your chosen topic. Select two measurement variables in the data file of interest to you. Use Excel to calculate the averages for those variables using the AVERAGE function. Discuss what the averages mean in terms of the variables and topic of interest.

5. Referring to Q4, the averages calculated for the two variables were population averages. Use a random number generator (www.randomizer.org/) to generate a random sample of size 20 from the two variables in your Excel file—the population data. Generate 20 random numbers from 1 to the last row of the Excel file—each number will represent a row in the Excel file. Select the measurement values for both variables from the 20 row numbers selected. Calculate the sample averages for both variables. Repeat this process. Compare the two sample averages for both variables to the population averages.

CHAPTER 2

ASSESSING THE QUALITY OF RESEARCH

Finding the occasional straw of truth awash in a great ocean of confusion and bamboozle requires intelligence, vigilance, dedication and courage. But if we don't practice these tough habits of thought, we cannot hope to solve the truly serious problems that face us and we risk becoming a nation of suckers, up for grabs by the next charlatan that comes along.

—Carl Sagan

2.1 INTRODUCTION

We are living in the Information Age. With countless sources of news, how do we decide what to believe and what not to believe? We can trust our favorite media source and assume that they are doing a good job of fact-checking the information they are presenting. As an alternative, we can learn to critically assess the quality of the research behind the news headlines for ourselves.

A story reported by CBS News titled "How the Chocolate Hoax Fooled Millions" is a good example of how relying on the media to assess the quality of research could lead you to believing nonsensical conclusions. A science journalist named John Bohannon with a PhD in molecular biology, decided to show how easy it is to turn a poorly conducted piece of research into news headlines.

KEY TERMS

Hypothesis Testing: A method of statistical analysis used for making statistical decisions regarding the values of population parameters

Null Hypothesis: A general statement or default position regarding the value of a population parameter. When comparing treatments, it is a statement that the treatments are equally effective

Alternative (Research) Hypothesis: A statement rejecting the null hypothesis. When comparing treatments, it is a statement that the treatments are not equally effective. The alternative hypothesis is our research hypothesis

Valid Measurement: Measures what it is supposed to measure

Reliable Measurement: Instrument used consistently measures what it is supposed to measure

Biased Measurement: The error in measurement is in one direction only favoring a particular result

Measurement Error: The difference between the measured value and the true value

Baseline: A measurement taken at the beginning of the study

Hazard Ratio: A statistic that compares the (relative) rate of a particular health outcome (over time) for one group compared (or relative) to another group

Statistical Decision: A decision based in chance regarding the value of a population parameter

p-value: Non-technically, it is the probability that the difference (in the sample estimates of population parameters) we observed between treatment groups is due to sampling variation alone

Level of Significance: A borderline value used for making a statistical decision. It is generally accepted that when a p-value is less than 0.05, it is considered statistically significant

Sample Effect Size: When comparing treatment groups, it is the size of the difference in sample statistics between treatment groups (e.g. average cholesterol levels)

Blinded: When a participant or researcher do not know which treatment the participant is on

Placebo Effect: The response of a patient to a treatment (both for placebo or real treatment) due to the patient's belief that the treatment will work

Prospective Study: A type of observational study. Participants are studied over a period of time and measurements for particular health outcomes (for example) are taken

Meta-Analysis: Combing multiple studies together to increase the power of the resulting analysis, and to obtain better estimates of the population parameters of interest

Cross-Sectional Study: A type of observational study that analyzes data from a population at a particular point in time

HOW THE CHOCOLATE HOAX FOOLED MILLIONS

Bohannon and the filmmakers concocted a plan to prove just how easy it is to turn bad science into big headlines. They created a website for the Institute of Diet and Health (a group they made up), recruited a doctor and analyst, and paid research subjects to take part in a small clinical trial they would run to test the effects of eating chocolate. Then Bohannon would use his media savvy to get the results published and publicized.

Web Link: http://www.cbsnews.com/news/how-the-chocolate-diet-hoax-fooled-millions/

Search Term: How the Chocolate Hoax Fooled Millions, CBS News

The results of the study found that eating chocolate every day can help you lose weight. Bohannon discusses how they came to this conclusion by manipulating the way in which we make **statistical decisions** regarding the relationships between variables using **hypothesis testing**. A statistical decision is a decision based in chance. If we look for relationships between many different variables, as this study did, we will eventually find a relationship between two variables that is statistically significant by chance alone. In Integrity in Research, the final chapter of this book, we will discuss in depth how this sort of misuse of the statistical method can and does occur.

Research is difficult. Going from a question of interest about the world around you to setting up a study and collecting measurements for analysis is easier said than done. Some research ideas are easier to implement than others, and some researchers do a better job than others. There are many ways errors or biases can be an inherent part of a study, affecting the measurements taken and the validity of the results.

In this chapter, our aim is to learn how to critique the quality of research for ourselves by asking the following questions:

- Who funded the study?
- What motivated the participants to be part of the study?
- What were the motivations of the researchers?
- How did the researchers and participants interact?
- Was the study a randomized experiment or an observational study?
- How were the participants selected?

- How well did the researchers measure what they wanted to measure?
- What is the practical significance of the results of the study?

The aim of the researcher should be to obtain a sample statistic that is a good estimate of the population parameter of interest. It is important to question the study design in order to determine any faults in the design that could lead to error or bias in the measurements taken. As stated at the end of the first chapter, our aim is to minimize the distance between our estimate of truth and the truth itself, by minimizing error and eliminating bias in our measurements.

There are a lot of research results presented in the media that are just plain bad science. Oftentimes, the results are based on observational data, where cause-and-effect conclusions should not be made. However, this does not prevent the researcher (and in turn the media) from making such claims. *Last Week Tonight* with John Oliver did a very entertaining and educational piece titled "Scientific Studies" on the types of research often presented in the media. As Oliver points out, there are so many scientific studies reported in the media with nonsensical and contradictory results. He talks about how scientists are under extreme pressure to publish as often as possible. As a result, there are many ways in which a scientist may "consciously or not tweak their studies".

LAST WEEK TONIGHT WITH JOHN OLIVER: "SCIENTIFIC STUDIES"

Web Link: https://www.youtube.com/watch?v=0Rnq1NpHdmw

Search Term: Last Week Tonight with John Oliver: Scientific Studies

In the following examples, we will see how problems with study design, sample selection and measurement can lead to poor quality research. In our final chapter titled Integrity in Research, we will discuss how dishonesty and a lack of integrity can also affect the quality of the research.

No piece of research is perfect. As we will learn throughout this book, the path to truth in science is steep with many potential pitfalls along the way. It is a challenging task for the researcher to ensure all types of error and bias (intentional or not) that could be an inherent part of the data are avoided. By going beyond the news headlines to critique the journal article for ourselves, we will begin to develop our critical and statistical thinking skills and gain a better understanding of how challenging it is to conduct good research.

2.2 ALCOHOL CONSUMPTION AND CARDIOVASCULAR HEALTH

In 2018, there were numerous news headlines related to the dangers of alcohol consumption. The news articles were based on two very large observational studies. What one might conclude from the news headlines is that the safest amount of alcohol to consume is none at all.

NEWS ARTICLE: NO ALCOHOL SAFE TO DRINK, GLOBAL STUDY CONFIRMS

Web Link: https://www.bbc.com/news/health-45283401

Search Term: No alcohol safe to drink, global study confirms

NEWS ARTICLE: HOW MUCH ALCOHOL IS SAFE TO DRINK? NONE, SAY THESE RESEARCHERS

Web Link: https://www.nytimes.com/2018/08/27/health/alcohol-drinking-health.html

Search Term: How Much Alcohol Is Safe to Drink?

RISK THRESHOLDS FOR ALCOHOL CONSUMPTION

Risk thresholds for alcohol consumption: combined analysis of individual-participant data for 599,912 current drinkers in 83 prospective studies

Web Link: https://www.thelancet.com/journals/lancet/article/PIIS0140-6736(18)30134-X/fulltext

Search Term: Risk thresholds for alcohol consumption

We will go beyond the news headlines to critique one of the two pieces of research for ourselves.

The results of the piece of research we will critique were published in *The Lancet* in a journal article titled "Risk thresholds for alcohol consumption: combined analysis of individual-participant data for 599,912 current drinkers in 83 prospective studies." A **prospective study** is a type of observational study. Participants are studied over a period of time, and measurements for particular outcomes (health-related, for example) are taken. When numerous studies are combined, the resulting analysis of the combined studies is known as a **meta-analysis**.

Meta-analysis can be a powerful way of getting at the truth in populations. It makes sense to believe that the more data we have, the closer our resulting analysis of that data will be to the truth. However, it depends on the type of studies and the type of data we are combining. For example, combining the data from numerous, very similarly designed randomized experiments with the same quantitative measurements can be a very powerful way of getting at the truth. However, combining the results of numerous observational studies using self-reported measurements, measured in different ways, is not nearly as powerful.

The alcohol study brought together 83 prospective studies resulting in data from 599,912 drinkers. At the beginning of each study selected, the drinkers had no history of cardiovascular disease. The researchers were interested in looking at the relationship between the explanatory variable: the amount of alcohol consumed and various health outcomes: stroke; coronary disease excluding myocardial infarction; heart failure; fatal hypertensive disease; fatal aortic aneurysm; and myocardial infarction.

The results of any research analysis are built upon data. The first question we need to ask is how did the researchers measure what they wanted to measure. Are the measurements taken valid, reliable, and unbiased? In other words, do they accurately measure what they are supposed to measure? A measurement is **valid** if a suitable instrument is used to take the measurement; a weighing scale is a valid way of measuring a person's weight. A measurement is **reliable** if the instrument used results in a consistent and accurate measurement when used repeatedly on the same person or object. A measurement is **biased** if there is a tendency for the instrument used to provide measurement values either above or below the true value. In other words, bias is error in your measurement in one direction only—either above or below the true value. For example, a weighing scale that provides a weight measurement 2 lbs. below the true weight every time it is used is considered a biased measurement. Ideally, we want to attain a measurement value that is equal to the true value. This is not easy to attain even for the most quantitative of measurements. Researchers should always ask themselves the question: Are we truly measuring what we want to measure?

In the case of a meta-analysis, we also want to know if the measurements were consistently taken across all studies. If not, how did the researchers take account of this issue? In the journal article, the researchers state that:

> *We did a combined analysis of individual-participant data from three large-scale data sources available to our consortium, each constituting purpose-designed prospective cohort studies with quantitative information about alcohol consumption (Appendix p. 21).*

The appendix gives greater detail on how the researchers went about "harmonizing" data on alcohol consumption across the 83 prospective studies. The data on alcohol consumption was collected in various ways across the studies. Examples were self-administered questionnaires, interview-led questionnaires, food-frequency questionnaires, and dietary recall surveys and questionnaires. The researchers did an exhaustive job to ensure the alcohol consumption and the cardiovascular disease event measurements were as similar as possible.

> *Therefore, our combined analysis included information from a total of 83 prospective studies that each used broadly similar methods to quantify alcohol consumption, record risk factors, and ascertain cause specific death and cardiovascular disease events.*

However, it must be understood that no matter how hard the researchers worked to harmonize the data, it would be impossible to remove all the error or bias contained in the measurements. Most of the data on alcohol consumption was collected in the form of questionnaires. Whether self-administered or interview-led, the quality of the resulting data is questionable. When asked how much one drinks on average per week, for many people, their estimates may deviate somewhat from the truth. How much one drinks from week to week may vary. When asked how much one drinks on average per week, an individual is not necessarily going to give an accurate estimate. Also, there is a certain stigma related to drinking too much. Therefore, many heavy drinkers might purposely underestimate the amount they drink on average each week. In the analysis of the data, these individuals would be categorized as, say, moderate drinkers instead of heavy drinkers. If a significant proportion of these drinkers were found to have cardiovascular events, the analysis results would overstate the dangers of moderate drinking as it relates to cardiovascular health.

Also, the individuals in this study (from 19 high-income countries) had varying tastes with regard to the type of alcohol they drank. A list provided in the appendix included cider, wine, sweet liquor, red wine, white wine, champagne, beer, spirits, alcopops, Finnish long drink, sake, shochu, and tharra. The researchers converted the varying measurement values to a standard measurement used for analysis (1 unit = 8 grams of ethanol). Again, you can imagine this is a very difficult task, prone to error or bias in the standardized measurements.

One final point to make with regard to the alcohol measurement used is that (in the appendix) the researchers present a table (eTable 4) showing that only 37 of the 83 combined studies had repeat alcohol measurements. In other words, for most of the studies used, the average amount of alcohol consumed per week was based on one measurement taken at one point in time. Since the participants were followed into the future (and any cardiovascular events recorded), one would have to assume that the average amount of alcohol consumed per participant would change over time.

With regard to cardiovascular event measurements, the researchers state that the various studies selected used broadly similar methods for assessing cardiovascular events but don't go into much further detail in either the journal article or the appendix. The fact that the types of measurements taken (coronary disease, myocardial infarction, heart failure, fatal hypertensive disease, and fatal aortic aneurysm) result in yes/no type answers, we have to assume the resulting measurements were accurate.

Besides looking at the relationship between alcohol consumption and cardiovascular events (the results of which we will discuss), the researchers also looked at the relationship between alcohol consumption and life expectancy:

In comparison to those who reported drinking >0–≤100 g per week, those who reported drinking >100–≤200 g per week, >200–≤350 g per week, or >350 g per week had lower life expectancy at age 40 years of approximately 6 months, 1–2 years, or 4–5 years, respectively.

For those who drank between >0–≤100 g per week, the average was 53 grams where a 100 g is equivalent to 5 or 6 pints of beer or glasses of wine. The statistics presented on life expectancy can seem alarming, suggesting that consuming more than one drink per day will shorten your life expectancy. However, it must be understood that these statistics are averages based on data from a large number of individuals from 83 observational studies where the measurements for alcohol consumption are questionable. As already mentioned, due to the nature of how the data was collected, many of the chronic drinkers (>350 g per week), for example, may have been categorized *as moderate drinkers (>100–≤200 g per week)*, adversely affecting the predictions of life expectancy for moderate drinkers.

However, even if there were no problems with the measurements taken, the other issue is that the results are based on observational data for which potential confounding factors need to be controlled. The researchers did control for particular confounding factors they had information on for all studies included in the analysis. In the summary of their findings, the researchers state that they controlled for age, sex, history of diabetes, and smoking status. On further reading of the content of the journal article, the researchers stated that they controlled for other confounding factors for individuals where the information was available. These included body-mass index (BMI), blood pressure, cholesterol levels, education, occupation, and self-reported general health. However, they do not clearly state (in the journal article or the appendix) how many individuals this information was available for and why these confounding factors were not controlled for in the primary analysis.

The researcher's efforts to control for confounding factors would certainly help them see more clearly the relationship between alcohol consumption, cardiovascular health, and life expectancy. However, despite their efforts, the results are questionable and causal conclusions can't be made. As discussed in Chapter 1, the nature of observational data is where there are many factors (known and unknown) that are related to (or confounded with) the explanatory variable that may affect the response variable. This fact makes it impossible for us to make cause-and-effect conclusions. There will always be confounding factors that were not accounted for.

Every individual who drinks (and the vast majority of people drink) may lead very different lives. For example, a 35-year-old male, with no history of diabetes and who smokes, may be considered a heavy drinker but also has poor diet, does not exercise, is slender, is depressed, does not sleep well, suffers from high anxiety, does not rehydrate after drinking, sits inside most of the day, drinks alone, lives alone, and has no friends or family. He drinks to numb the pain of his existence. Many of these factors would adversely affect his cardiovascular health and life expectancy.

Another 35-year-old male, with no history of diabetes and who smokes, may be considered a heavy drinker but eats healthily, exercises every day, is stocky, is very happy, gets a restful sleep every night, has little to no anxiety, rehydrates after drinking, is active all day long, only drinks socially, has many friends and family, and lives with a wife and children that he loves and adores. He drinks because he enjoys it and it takes the edge off at the end of the day along with great laughs and conversations with friends and family. He enjoys life and is in complete control. Many of these factors would positively affect his cardiovascular health and life expectancy.

Of course, this is an extreme example in order to make a point. In their primary analysis, the researchers may have controlled for age, sex, history of diabetes, and smoking status. However, if we take two individuals who drink heavily with the same values for these confounding factors, they may differ considerably on other confounding factors (not accounted for in the primary analysis) that affect their cardiovascular health and life expectancy in very different ways. The point is that the world of observational data is complex because each individual lives his/her life differently. There are numerous factors that affect our cardiovascular health and life expectancy, and they are not necessarily easy to measure or control for in an observational study.

There are other questions we could ask about this study: who funded the study, what are the motivations of the researchers and participants, how did they interact with each other, and how were the participants selected? This study was a meta-analysis; to determine the motivations of the participants, how they were selected, and how they interacted with the (original) researchers, we would have to find and critique the journal articles related to the original 83 studies. The purpose of this discussion is to critique the meta-analysis study and not the individual studies.

For this study, we can ask about funding and the motivations of the researchers involved. From reading the journal article, there were a large number of researchers involved in this research with grants from numerous sources, including UK Medical Research Council, British Heart Foundation, and the National Institute for Health Research, to name a few. Looking at the long list of respected researchers and funders, one has to assume that this research was a sincere attempt to access the effect of increasing levels of alcohol consumption on particular health outcomes.

Finally, we will discuss the practical significance of the results relating alcohol consumption to cardiovascular health. The primary findings were presented as follows:

Alcohol consumption was roughly linearly associated with a higher risk of stroke (HR per 100 g per week higher consumption 1·14, 95% CI, 1·10–1·17), coronary disease excluding myocardial infarction (1·06, 1·00–1·11), heart failure (1·09, 1·03–1·15], fatal hypertensive disease (1·24, 1·15–1·33); and fatal aortic aneurysm (1·15, 1·03–1·28). By contrast, increased alcohol consumption was log linearly associated with a lower risk of myocardial infarction (HR 0·94, 0·91–0·97).

The researchers found a minimal mortality risk for drinkers who consumed approximately 100 g of alcohol per week or less. They calculated the change in risk (as compared to these drinkers on various health outcomes) for each *100 g per week* increase in alcohol consumption across drinkers. The statistic used for making these comparisons is called the **hazard ratio** (HR). This is simply a statistic that compares the (relative) rate of a particular health outcome (over time) for one group compared to (or relative to) another group. For example, when comparing two groups, if the second group experienced a particular health outcome at twice the rate of the first group, the hazard ratio would be equal to 2.

In this study, the **baseline** group were drinkers who consumed 100 g of alcohol per week or less. Each subsequent group of drinkers (incremented by 100 g of alcohol consumption per week) was compared to this baseline group of drinkers. For example, the hazard ratio for heart failure was 1.09. This means that for a 100 g increase in alcohol consumption per week, risk of heart failure increases by 9%. In order to understand what this 9% increased risk really means, we need to know what is the risk of heart failure for minimal mortality risk drinkers.

From reading the study, we can conclude that that the risk of heart failure for these drinkers was found to be approximately the same as it is for nondrinkers.

In *Science News*, an article titled "Drinking studies muddied the waters around the safety of alcohol use," the author discusses the results of both alcohol studies published in 2018. From this discussion, we can conclude that the increased risk of death from a heart condition as a result of alcohol consumption is very small:

> Study coauthor Emmanuela Gakidou, an expert in health metrics, acknowledges that the risks for light to moderate drinkers are small. In a given year, 914 per 100,000 people who drink no alcohol will die from one of the health conditions examined in the study. If all those people had one drink per day in that year, an extra four, for a total of 918, would die.

NEWS ARTICLE: DRINKING STUDIES MUDDIED THE WATERS AROUND THE SAFETY OF ALCOHOL USE

Web Link: https://www.sciencenews.org/article/alcohol-drinking-health-risk-top-science-stories-2018-yir

Search Term: Drinking studies muddied the waters

Since the data is observational in nature and the increased risks the researchers are trying to estimate are very small, it is very difficult to separate the actual increased risk of alcohol consumption from the other confounding factors that might increase or decrease the risk of the various health outcomes. It is interesting that the researchers found a lower risk of myocardial infarction (heart attack) from increased alcohol consumption with a hazard ratio equal to 0.94. Again, it is very difficult to know how much of this perceived decreased risk is due to alcohol consumption and how much is due to other factors, including measurement error.

The researcher Emmanuela Gakidou was also quoted as saying:

> "Saying to yourself, having a glass of wine presents a small risk, but I enjoy it—OK, that's fine. But I would like people to move away from thinking drinking is good for you."

This is a fair enough point to make. However, as already discussed, how adversely drinking affects your physical health depends on not only how much you drink but also on many other factors related to your physical makeup and how you live. From our critique of this study, we have learned that with observational data it is impossible to determine cause-and-effect relationships due to confounding factors, especially when the true effect size is small.

Much of the research we read about in the media is based on observational data. The news headlines may give the impression that cause-and-effect conclusions can be made from these studies. As we develop our statistical thinking, it is very important to understand that not all data is created equally. Measuring what we want to measure can be very difficult, and cause-and-effect conclusions can only be made from well conducted randomized experiments.

WITH BOTOX, LOOKING GOOD AND FEELING LESS

It's no shock that we can't tell what the Botoxed are feeling. But it turns out that people with frozen faces have little idea what we're feeling, either.

No, Botox injections don't zap brain cells. (At least not so far as we know.) According to a new study by David T. Neal, an assistant professor of psychology at the University of Southern California, and Tanya L. Chartrand, a professor of marketing and psychology at the Duke University Fuqua School of Business, people who have had Botox injections are physically unable to mimic emotions of others. This failure to mirror the faces of those they are watching or talking to robs them of the ability to understand what people are feeling, the study says.

Web Link: http://www.nytimes.com/2011/06/19/fashion/botox-reduces-the-ability-to-empathize-study-says.html

Search Term: With Botox, Looking Good and Feeling Less

EMBODIED EMOTION PERCEPTION: AMPLIFYING AND DAMPENING FACIAL FEEDBACK MODULATES EMOTION PERCEPTION ACCURACY

Web Link: http://spp.sagepub.com/content/2/6/673

Search Term: Embodied Emotion Perception

2.3 WITH BOTOX, LOOKING GOOD AND FEELING LESS

The second piece of research we will critique relates to the effects of Botox on a person's ability to pick up on the emotions of others. The *New York Times*, in an article titled "With Botox, Looking Good and Feeling Less", reported on the research without questioning the validity of the research in any way.

As the news article points out, the researchers in this study were interested in how facially administered Botox affects a person's ability to pick up on the emotions expressed by others. It is an interesting hypothesis. The idea is that when we are engaged in a face-to-face conversation with a friend or colleague, we tend to mimic their facial expressions so that we can better empathize with the feelings or emotions they are trying to express. Oftentimes, if we observe two people (usually friends or family) engaged in a conversation, this is what it looks like they are doing.

The results of the research were presented in the journal *Social Psychology and Personality Science* in a journal article titled "Embodied Emotion Perception: Amplifying and Dampening Facial Feedback Modulates Emotion Perception Accuracy."

The researchers completed two experiments. We will focus on Experiment 1. To analyze the data, the researchers conducted a hypothesis test. In this experiment, the **null hypothesis** is that emotion perception accuracy of the perceiver is unaffected by the use of Botox. The researcher's **alternative (research) hypothesis** is that Botox dampens emotion perception accuracy. We can think of the null hypothesis as analogous to the presumption of innocence at the beginning of a jury trial. The alternative hypothesis is analogous to the rejection of innocence in favor of a guilty verdict. Only when the statistical evidence (found in our data) is beyond a reasonable doubt will we reject the null hypothesis in favor of the alternative.

We will begin our critique by focusing on the type of measurement taken. In Experiment 1, emotion perception accuracy of the individuals selected for the study is measured using what is known as Reading the Mind in the Eyes Test (RMET):

On each RMET trial, a black and a white photograph appeared at central fixation. The photograph depicted only the eyes and immediate

surrounding area. Four adjectives appeared simultaneously with the image: One adjective represented the correct answer and the other three were foils typically of the same valence as the correct response. Participants were instructed to select the emotion that best matched the expression and to maximize accuracy and speed. One practice trial was given prior to the 36 test trials.

If you are curious about how this test works, you can try it out for yourself. An article in the *New York Times* titled "Can You Read People's Emotions?" contains a link to the test.

This method of assessing a person's ability to pick up on the emotions of others would definitely help keep the costs of the research down. However, it has its limitations. The perceiver is limited to four emotions to choose from, making it easier to make the right choice by a process of elimination. A perceiver who is good at multiple-choice questions has an advantage. Why just the eyes and not the whole face? Certainly, when we look in the mirror and express different types of emotions, we notice that the emotion is most apparent in the eyes, but the lower part of the face is undoubtedly a part of the emotional expression, too. In reality, a person perceives the whole face and therefore the complete emotion expressed. Thus, would it not be more realistic to present the entire face? Perhaps they could have used an actual person, say an actor, expressing the emotions face-to-face with the perceiver. Looking at a black and white photograph of eyes on a screen is not the same thing as as having the actual person in the room. It would be a more costly way of measuring a person's ability to pick on the emotions of others, but it would lead to more accurate measurements.

CAN YOU READ PEOPLE'S EMOTIONS?

Web Link: http://well.blogs.nytimes.com/2013/10/03/well-quiz-the-mind-behind-the-eyes/

Search Term: Can You Read People's Emotions?

Next, we will critique how the study was designed. Patients receiving either Botox or a dermal filler called Restylane were recruited for the study:

> *Experiment 1 tested whether emotion perception accuracy declines when perceivers' facial feedback has been dampened. To do this, we recruited a matched sample of patients receiving either Botox injections or Restylane injections for the cosmetic treatment of expressive facial wrinkles. Botox paralyzes the expressive muscle through blocking acetylcholine release at the neuromuscular junction (Dolly & Aoki, 2006). Critically, this paralysis reduces afferent feedback from the injected muscles to the brain (Hennenlotter et al., 2009). Restylane is a dermal filler and does not alter muscle function, thus leaving facial feedback intact (Brandt & Cazzaniga, 2007).*

The researchers do not explain what they mean by a matched sample of patients. This statement gives the impression that the patients were similar in every way besides the treatment they were on. However, this is an observational study and not a randomized experiment. The patients chose their preferred treatment for expressive facial wrinkles. The researchers make the added claim that prior research has shown that both groups have the same baseline response to emotional stimuli:

> *The two patient groups did not differ in mean age, ethnicity, or socioeconomic status. In addition, although random assignment was not possible given the clinical setting, prior research shows that patients receiving Botox and Restylane do not differ from one another in their baseline reactivity to emotional stimuli prior to treatment.*

This is quite a strong claim to make. It implies that if the researchers find the results to be **statistically significant** (which they did), then they could also make a cause-and-effect conclusion. However, there could be other factors that differ between the two groups of patients that affect their ability to pick up on other people's emotions. Maybe some of the patients who choose Botox are less empathetic on average than patients who choose Restylane. Maybe there is a higher proportion of patients in the Restylane group who are not good at tests like the RMET. Maybe the opposite is true. This was not a randomized experiment so we have to allow for the fact there could be many confounding factors affecting the response, besides the treatment the patients were on.

Regardless of whether or not prior research has shown no difference between the two groups' baseline response to emotional stimuli, why not measure the baseline response to be sure? A change from baseline to endpoint measurement could be used in the final analysis, strengthening the quality of the data collected and the research results.

We now want to think about how the patients were recruited for the study. Ideally, the researcher should clearly define the population of individuals and select a representative sample from that population. According to the researchers, the sample consisted of 31 females with a mean age of 52 years. The 31 females were recruited from cosmetic surgery clinics. A sample of patients recruited in this way is considered a convenience sample. As discussed in Chapter 1, many studies resort to convenience samples because of the difficulties in choosing a random sample of patients. The question is how representative is this convenience sample of the population? To what population can the results based on this sample be extended?

We are told the patients "were recruited from cosmetic surgery clinics and paid $200 for completing the study," but we are not told which clinics and where they were located. We have to assume that the researchers would like to extend their results to the population of women in the United States who attend cosmetic surgery clinics for the treatment of expression-related wrinkles. In order to do so, a random sample of women from all such clinics in the United States needed to be selected. At best, the results could be extended to all of the patients who attend the clinics the sample of patients were recruited from.

However, the sample of patients were paid $200 for participating in the study. The patients who were enticed by the $200 to be part of the study may differ in certain characteristics from those who decided not to be part of the study. Whether or not they are a representative sample of patients depends on how much they differ in these characteristics from the entire population of women who attend such clinics. It is difficult for us to determine what these characteristics were. Perhaps lower income patients were more likely to participate in the study. If so, are lower income patients more or less likely than high income patients to pick up on the emotions of others, regardless of what treatment they were on? Maybe or maybe not. If it turned out that the subset of patients that participated in the study were all low-income patients, then the results of the study should only be extended to that population of patients. When researchers rely on convenience samples, it can be very difficult to know what population the results can be extended to.

Another possible source of bias is how the researcher interacted with the patients. A researcher is looking for evidence in favor of their hypotheses. To eliminate any bias on the part of the researcher, they should be **blinded** to what treatment the patient is on throughout the entire study. Ideally, the researchers should have a separate person administer the tests and collect the data, keeping the main researcher blinded to what treatment the patient is on throughout the process.

In this study, we are not told how the researcher and patients interacted. If the main researcher recruited the patients and administered the tests, then this could be a problem. The researcher may (consciously or unconsciously) bias the results in some way. If at all possible, the main researcher should be one step removed from the process of collecting the data.

Let's now look at the results of Experiment 1 to discuss how significant the results were. The sample statistics used in this study were the mean percentage of times the patients chose the correct emotion in the Reading the Mind in the Eyes Test (RMET) for both groups. This percentage was 69.91% for the Botox group and 76.92% for the Restylane group. The difference of 7% between these sample statistics was found to be statistically significant because the **p-value** equal to 0.046 is less than 0.05, known as the **level of significance**. It is a borderline value used for making a statistical decision that we will discuss in depth in a later chapter. The small p-value means that the researchers found statistical evidence of a difference in the population in the ability of these two groups to pick up on the emotions of others.

However, even if the difference of 7% (known as the **sample effect size**) between the sample statistics is an accurate estimate of the real difference in the population, what is the practical significance of the result? Yes, the result suggests that Botox users are not as good as Restylane users at picking up on the emotions of others, but they still got the answer right 70% of the time on average. The Restylane users did not do that much better, getting the answer right 77% of the time.

The use of the word *significance* when describing the results of a study is often misinterpreted or misused, suggesting that the research findings are more important than they actually are. A statistical decision regarding the true value of the effect size is one based in chance. A statistically significant result simply means the effect size we observe in our sample data (or one even larger) is unlikely to have happened by chance, given that there is no effect in the population. Whether or not the effect size found is of practical significance depends on the particular research question and the measurement taken.

Finally, we will look at who funded the study and how it may have affected the motivations of the researchers to find a positive result. The journal article states that:

> The author(s) received no financial support for the research, authorship, and/or publication of this article.

What were the researchers' motivations for completing the research if the researchers had to completely self-finance the study? Why were they interested in showing that Botox dampens people's emotion perception accuracy? What were the benefits to the researchers of finding a statistically significant result? The journal article does not provide answers to these questions.

This piece of research was an observational study. This means that even though the results were statistically significant, we can't make causal conclusions. We can't say that the Botox injections causes people's emotion perception accuracy to decline. Even if we could make a causal conclusion, the difference in outcomes found between the Botox and Restylane patients probably has no real practical significance. In other words, it is unlikely to stop someone from getting their Botox injections, out of fear of losing their ability to pick up on the emotions of others.

2.4 THE FAST DIET

With obesity rates at an all-time high in the United States, losing weight has become an important issue for many Americans. It has also become big business with the emergence of reality shows like *The Biggest Loser* and many different types of diets that claim to be effective.

When people decide to lose weight, oftentimes they want to lose it as fast they can. In the *Time* magazine article titled "The Weight Loss Trap: Why Your Diet Isn't Working" a scientist named Kevin Hall discusses

---

THE WEIGHT LOSS TRAP: WHY YOUR DIET ISN'T WORKING

Hall, a scientist at the National Institutes of Health (NIH), started watching The Biggest Loser a few years ago on the recommendation of a friend. "I saw these folks stepping on scales, and they lost 20 lb. in a week," he says.

...

Over the course of the season, the contestants lost an average of 127 lb. each and about 64% of their body fat. If his study could uncover what was happening in their bodies on a physiological level, he thought, maybe he'd be able to help the staggering 71% of American adults who are overweight.

What he didn't expect to learn was that even when the conditions for weight loss are TV-perfect—with a tough but motivating trainer, telegenic doctors, strict meal plans and killer workouts—the body will, in the long run, fight like hell to get that fat back. Over time, 13 of the 14 contestants Hall studied gained, on average, 66% of the weight they'd lost on the show, and four were heavier than they were before the competition.

Web Link: http://time.com/4793832/the-weight-loss-trap/
Search Term: The Weight Loss Trap: Why Your Diet Isn't Working

the results of his research into the contestants on *The Biggest Loser* and the difficulties they had in maintaining their weight loss.

The article discusses the many challenges people face when trying to lose weight and the rise in obesity research to help meet those challenges. What scientists have found is that while exercise is good, it is not the most reliable way to lose weight. Also, a personalized diet is a better way to lose weight than following the latest dietary trends.

One type of diet growing in popularity is known as the 5:2 diet. The diet entails fasting for two days a week and eating normally for the other five days. As the *New York Times* article titled "Fasting Diets Are Gaining Acceptance" points out, the 5:2 diet has become very popular with celebrities like Benedict Cumberbatch, Hugh Jackman, and Jimmy Kimmel.

The 5:2 diet has also been promoted in several books. One of those books, *The Fast Diet* by Dr. Michael Mosley, is discussed in the *New York Times* article titled "England Develops a Voracious Appetite For a New Diet".

The article points out that Mosley discusses the science behind the diet in the first two hundred pages of the book, but the article's author does not question the validity of the research that Mosley uses to back up his claims. The author does mention that the National Health Service in England put out a statement regarding this type of diet—also known as the Intermittent Fasting (IF) diet—saying that "there is a great deal of uncertainty about IF with significant gaps in the evidence."

We will go beyond the news headlines by going to the source of the research that Mosley used to make his claims regarding the benefits of the diet. After reading Mosley's book, we find that most of the benefits discussed in his book are based on the findings of one published piece of research. We will begin our critique of the research by comparing Mosley's claims about the benefits of the 5:2 diet to the research results. We will then focus our questions on the quality of the research.

FASTING DIETS ARE GAINING ACCEPTANCE

It has been promoted in best-selling books and endorsed by celebrities like the actors Hugh Jackman and Benedict Cumberbatch. The late-night talk show host Jimmy Kimmel claims that for the past two years he has followed an intermittent fasting program known as the 5:2 diet, which entails normal eating for five days and fasting for two—a practice Mr. Kimmel credits for this significant weight loss.

Web Link: http://well.blogs.nytimes.com/2016/03/07/intermittent-fasting-diets-are-gaining-acceptance/

Search Term: Fasting Diets Are Gaining Acceptance

ENGLAND DEVELOPS A VORACIOUS APPETITE FOR A NEW DIET

Visitors to England right now, be warned. The big topic on people's minds—from cabdrivers to corporate executives—is not Kate Middleton's increasingly visible baby bump (though the craze does involve the size of one's waistline), but rather a best-selling diet book that has sent the British into a fasting frenzy.

"The Fast Diet," published in mid-January in Britain, could do the same in the United States if Americans eat it up. The United States edition arrived last week.

The book has held the No. 1 slot on Amazon's British site nearly every day since its publication in January, according to Rebecca Nicolson, a founder of Short Books, the independent publishing company behind the sensation. "It is selling," she said, "like hot cakes," which coincidentally are something one can actually eat on this revolutionary diet.

Web Link: http://www.nytimes.com/2013/03/03/fashion/england-develops-a-voracious-appetite-for-a-new-diet.html

Search Term: England Develops a Voracious Appetite for a New Diet

THE EFFECTS OF INTERMITTENT OR CONTINUOUS ENERGY RESTRICTION ON WEIGHT LOSS AND METABOLIC DISEASE RISK MARKERS: A RANDOMIZED TRIAL IN YOUNG OVERWEIGHT WOMEN

Web Link: http://www.nature.com/ijo/journal/v35/n5/full/ijo2010171a.html

Search Term: The Effects of Intermittent or Continuous Energy Restriction

The research results were presented in the *International Journal of Obesity* in a journal article titled "The Effects of Intermittent or Continuous Energy Restriction on Weight Loss and Metabolic Disease Risk Markers: A Randomized Trial in Young Overweight Women."

The study was a randomized experiment. The participants were randomized to one of two diets: the intermittent energy restriction (IER) diet or the continuous energy restriction (CER) diet. The IER diet group were allowed to consume 2710 kJ (or 650 calories) per day for two days a week and eat normally on the other five days. The CER diet group were allowed to consume 6276 kJ (or 1500 calories) per day, seven days a week.

In his book, Dr. Mosley states that the participants on the IER diet lost more weight on average than the participants on the CER diet. However, he fails to mention that the difference in average weight loss between the two groups was not found to be statistically significant.

According to the study results, the average weight loss was 6.4 kg for IER dieters versus 5.6 kg for the CER dieters, a difference of 0.8 kg. With a p-value equal to 0.40—far from less than the borderline value of 0.05 required for statistical significance—there is little evidence in this study to suggest that the IER diet is more effective than regular dieting for weight loss. However, Dr. Mosley makes several other claims about the many benefits of the diet beyond weight loss.

In his book, Mosley states that for the women in this study:

> ... their fasting insulin and insulin resistance had fallen further, and levels of inflammatory protein were also significantly down. All three measures suggest a reduced breast cancer risk.

However, according to the journal article:

> Both groups experienced modest declines in fasting serum insulin and improvements in insulin sensitivity, which were greater among the IER group. ...

> Both groups experienced modest decreases in the inflammatory marker high-sensitivity C-reactive protein, but no change in sialic acid levels.

This is a good example of the author playing with the use of the word *significant*. The changes in fasting insulin and insulin resistance were statistically significant, both with p-values equal to 0.04. However, the journal article also states that the changes were modest and the biological significance of the changes are unknown. Changes in the inflammatory protein were also found to be modest in both groups of dieters, and nowhere in the journal article does it mention that the changes were statistically significant.

Mosley goes on to make claims about how fasting will reduce levels of a cellular stimulant known as IGF-1 and therefore lower your risk of cancer.

In his book, Mosley states:

The cells in our body are constantly multiplying, replacing dead, worn out and damaged tissue. This is fine as long as cellular growth is under control, but sometimes a cell mutates, and grows uncontrollably, and turns into cancer. Very high levels of cellular stimulant like IGF-1 in the blood are likely to increase the chances of this happening. ... Fasting, either prolonged or intermittent, will reduce your IGF-1 levels, and therefore your risk of different cancers.

However, according to the journal article:

Neither IER nor CER led to appreciable changes in total or free IGF-1. Animal studies have shown reductions in IGF-1 with CER, but not consistently with IER.

We will now focus on assessing the quality of the research in the journal article. We will begin by questioning how representative the selected sample was. The journal article states:

We studied 107 premenopausal women aged 30–45 years with adult weight gain exceeding 10 kg since the age of 20 years, and a body mass index between 24 and 40 kg m. We recruited women from our Breast Cancer Family History Clinic, and women from the general population. As such, 54% of recruits had a family history of breast cancer.

The journal article goes on to say that they recruited participants from the breast cancer clinic through mail and women from the general population through the media and e-mail.

The women who decided to participate in the study are not necessarily representative of all overweight women who attend the Breast Cancer Family History Clinic or all overweight women in the general population. They had their own motivations for participating, making them a very select sample of individuals.

This is the problem of using a convenience sample rather than a random sample. It is very difficult to select a truly random sample from a population. However, the further the sample used is from random, the more difficult it is to know what population the results can be extended to. As a somewhat conservative rule of thumb, the more narrowly defined the sample selected, the more narrowly defined the population the study results can be extended to.

Another potential problem with the study is that the motivations of the participants (for being part of the study) could have also affected the results. The study was a randomized experiment, but for obvious reasons, the participants were aware of (not blinded to) what diet they were randomized to. If some of the participants were motivated to be part of the study, hoping to be on the IER diet, would they still have been as motivated if they were randomized to the CER diet? If not, could this have affected how well they adhered to the CER diet?

What about the motivations of the researchers? The participants could not have been blinded to what diet they were on, but what about the researchers? Whenever possible, in a randomized experiment, the main researchers should be blinded to what treatment the participants are on. This is to ensure that any bias on their part, either conscious or unconscious, does not find its way into the study and affect the measurements taken. The main researchers should have a third party, such as a technician, collect the data and return it to the main researchers with what is known as a dummy treatment allocation. This means that the main researchers would be blinded to the actual treatment the participants were on. Only at the end of the study would the blind be broken so the main researchers can see the true results with the proper treatment allocation.

In this study, one of the two dietitians who interacted with the patients and encouraged them to stick to their diets was Michelle Harvie, the main researcher on the study. The journal article does not give any indication that she was blinded to what diet the participants were on. There is another reason why her close interaction with the participants could be a problem. Her company, Genesis Appeal Manchester UK, was one of the main funders of the study. The company brought out a book, *The 2-Day Diet*, on February 14, 2013, citing this research as evidence that an intermittent diet works as well as a regular diet for weight loss. It would have been a good idea if she had remained blinded to which diet the participants were on. Blinding would have ensured that any motivations on her part as to how the study should turn out would not affect the study results.

Let's now focus on the measurements used and how well they were taken. The journal article states that measurements were taken at the beginning of the study and at one month, three months, and six months. The measurements taken included total body fat, weight, and systolic and diastolic blood pressure. All measurements were taken in the morning, after the participants had fasted for twelve hours. The participants were wearing light clothing when weight and body fat measurements were taken.

How do we know that the measurements were taken correctly? No further information was given, so we have to assume that they were. We should also ask whether all participants adhered to the twelve-hour fasting rule. The journal article states that adherence to the diets was assessed using food diaries after one month, three months, and six months and checked for completeness. However, no information was provided on how well the food diaries were filled out over the six-month period of the study. Adherence to providing the dietary records was also pretty low:

Weekly dietary records were available for 82 (76%) subjects at baseline, for 72 (67%) at 1 month, 65 (60%) at 3 months and 58 (54%) at 6 months.

If many of the participants did not provide proof that they stuck to their diets, why would their weight changes at one month, three months, and six months be included in the final analysis?

The journal article states that 107 participants entered the study at baseline, but only 89 participants completed the study due to withdrawals. Most of these withdrawals occurred within the first two months. The statistical analysis portion of the article states that the researchers used a last-observation-carried-forward analysis. This means that the last weight measurement recorded for the participants who withdrew was used for the final analysis at six months. We should question the validity of analyzing the changes in weight measurements at the end of the study for participants who withdrew from the study after one or two months.

Furthermore, adherence to both diets over the course of the study was very low:

Intention-to-treat analysis assuming that women who left the study or who did not complete food diaries did not adhere to the diets shows reported adherence to 2-day very low calorie diet (VLCD) among the IER group to be 63% at 1 month, 43% at 3 months and 44% at 6 months. ... The proportion of CER subjects who reported to adhere to the 25% CER was 46% at 1 month, 37% at 3 months and 32% at 6 months.

How can the researchers state that the results were based on weight measurements taken from 107 participants when adherence rates to both diets were so low? It completely throws into question the validity of the final analysis. If participants are not following their diets, why are the researchers including these participants in the final analysis?

Finally, we will look at one more issue with the study relating to the randomization process. In Table I on page 5 of the journal article, the two groups of subjects randomized to both diets are compared on many different baseline characteristics, including age, weight, ethnic origin, and employment. A statistical analysis was done to compare the sample statistics for these factors in both groups as a way of checking that the randomization process was successful. All p-values were greater than 0.05, meaning that any differences between the sample statistics (for each factor) when comparing both groups were most likely due to chance. It appears that the randomization process worked. However, one comparison discussed raises some concern:

The majority of subjects reported previous attempts at dieting (IER 92%, CER 78%), with a comparable number of previous attempts between the groups: IER 2.8 (2.1) and CER 2.4 (1.9) (P=0.29).

The researchers state that 92% of patients on the IER diet reported previous attempts at dieting, while 78% of patients in the CER did. No p-value is given for testing whether these percentages are statistically significantly different from each other. Regardless, the fact that a higher percentage of the IER dieters had previously dieted would give the IER group somewhat of an advantage. The randomization process should have been repeated.

As stated in Chapter 1, randomized experiments are considered the gold standard in research, but not all randomized experiments are created equally. The best experiments are randomized, comparative, double-blind, placebo-controlled studies. Blinding the subjects to what treatment they are on is not always ethical or possible. Blinding the main researchers from what treatment the subjects are on should be implemented whenever possible. Including a group of subjects on placebo gives us a control group to compare to. It also cancels out a very powerful effect known as the **placebo effect**. This is the response of a patient to a treatment due to the patient's belief that the treatment will work. If the patients are randomized and blinded to what treatment they are on, then the placebo effect should be averaged out, along with all other confounding factors. In other words, because the patients don't know whether they are on the treatment or the placebo, (and were randomized to treatment and placebo) the average placebo effect should be the same for both groups. Therefore, any difference found in the response when comparing both groups can be contributed to the treatment.

2.5 AMERICAN FOOTBALL AND BRAIN INJURY

There have been a number of news articles in the media regarding growing concern about the risks of brain injury from concussion for American football players. Aaron Hernandez was one of the youngest players found to have the brain degenerative disease known as chronic traumatic encephalopathy (CTE). The finding was particularly controversial because he was a convicted murderer and committed suicide when he was 27 years old in April 2017.

THE LATEST BRAIN STUDY EXAMINED 111 FORMER NFL PLAYERS. ONLY ONE DIDN'T HAVE CTE

Web Link: https://www.washingtonpost.com/sports/the-latest-brain-study-examined-111-former-nfl-players-only-one-didnt-have-cte/2017/07/25/835b49e4-70bc-11e7-8839-ec48ec4cae25_story.html?utm_term=.0024e208886f

Search Term: Washington Post Brain Study

The results of one study that got a lot of media attention looked at the relationship between playing football and CTE. The researchers examined the brains of 202 former players who played at all levels of the game, from college level to the NFL. The brains were donated by concerned families. They found 87% of the brains, 177 of them, had CTE. Out of the 111 players that played for the NFL, 110 were found to have CTE.

When we first read the results of this study, we feel that the researchers have found overwhelming evidence that playing football is extremely dangerous for the vast majority of players at all levels of the game. Playing the game can result in head trauma and concussions, so there is little doubt it is dangerous. However, the results of this study were based on a convenience sample.

Ann McKee, a neuropathologist, is quoted in the *Washington Post* article titled "The Latest Brain Study Examined 111 Former NFL Players. *Only one didn't have CTE.*" An expert on CTE diagnoses, she states:

> *"A family is much more likely to donate if they're concerned about their loved one—if they're exhibiting symptoms or signs that are concerning them, or if they died accidentally or especially if they committed suicide," she said. "It skews for accidental deaths, suicide and individuals with disabling or discomforting symptoms."*

> *While the study isn't focused on causality, McKee says it provides "overwhelming circumstantial evidence that CTE is linked to football."*

The researchers were well aware that they were working with a convenience sample and clearly state this fact in the journal article titled "Clinicopathological Evaluation of Chronic Traumatic Encephalopathy in Players of American Football" presented in the *Journal of the American Medical Association*. The question is how extreme a convenience sample is this particular sample of player's brains?

By way of explanation, let's imagine a hypothetical scenario where 1 in every 100 players in the population of NFL players develops CTE. In other words, the rate of the disease in players is 1%. Let's say that in its entire history, 10,000 players have played for the NFL. This means we would expect 1% of the players, or 100 players, to develop CTE. These 100 players are the only players who would exhibit any signs and symptoms of the disease. The signs and symptoms would cause their families to be concerned. In the most extreme situation, the only brains to be donated would be from families whose concerns were real and justified. The results of the study would estimate the disease rate at 100%, when it is in fact only 1%.

No one piece of research is perfect. Most studies have to rely on convenience samples. For example, a company wanting to test a drug for a certain disease can't select a random sample of people with the disease, call them up, and expect them to take part in their study. They have to rely on volunteers, a convenience sample.

What makes a good study is not just how well the researchers design the study in terms of avoiding the types of errors and biases we have discussed so far. The scientific process is one of experimentation; trial and error. A good study is also one where the researchers recognize its limitations and discusses those limitations in the journal article.

CLINICOPATHOLOGICAL EVALUATION OF CHRONIC TRAUMATIC ENCEPHALOPATHY IN PLAYERS OF AMERICAN FOOTBALL

Web Link: http://jamanetwork.com/journals/jama/article-abstract/2645104

Search Term: Clinicopathological Evaluation of Chronic Traumatic Encephalopathy

> **RELATIONSHIP OF COLLEGIATE FOOTBALL EXPERIENCE AND CONCUSSION WITH HIPPOCAMPAL VOLUME AND COGNITIVE OUTCOMES**
>
>
>
> Web Link: http://jamanetwork.com/journals/jama/fullarticle/1869211
>
> Search Term: Relationship of Collegiate Football Experience and Concussion

As an example, we will look at another study looking at the relationship between playing American football and brain injury. The journal article was also presented in the *Journal of the American Medical Association* titled "Relationship of Collegiate Football Experience and Concussion with Hippocampal Volume and Cognitive Outcomes."

The researchers compared college football players with a history of concussion to players without a history of concussion. The researchers also compared both groups to healthy controls—non-football players matched by age, sex, and education. There were twenty-five players in each group. The key measurements were hippocampal volume (a measure of cognitive ability) and reaction times. The researchers found that players with and without a history of concussion had smaller hippocampal volumes (on average) than the healthy controls:

> *The hippocampus is a brain region involved in regulating multiple cognitive and emotional processes affected by concussion and is particularly sensitive to moderate and severe traumatic brain injury (TBI). …*
>
> *In the left hemisphere, hippocampal volume was 14.1% smaller for athletes with no history of concussion and 23.8% smaller for athletes with a history of concussion relative to controls. … Similarly, right hippocampal volume was 16.7% smaller for athletes with no history of concussion and 25.6% smaller for athletes with a history of concussion relative to controls.*

They also found a relationship between baseline reaction times and number of years played. We will take a closer look at the results of this study in later chapters.

As the researchers state in the journal article, the study is a type of observational study known as a **cross-sectional study**. Therefore, we can't make causal conclusions regarding the relationships found in the study results. The researchers point this out when discussing the limitations of the study:

> *Limitations of this study warrant discussion. First, the cross-sectional design prevents inferences regarding causality and temporality about the relationship between years of football experience and hippocampus volume.*

They also point out that (because it is an observational study) there are many other factors that may result in differences in hippocampal volume across each group:

> *The interpretation of hippocampal volumetric differences is complicated by multiple factors including genetics, environment, hormones, growth factors, and neurodevelopmental trajectory. Among the*

most consistent findings associated with smaller hippocampal volumes are high levels of stress-related hormones. Collegiate athletes have been exposed to both physical and psychological stressors throughout their careers. These stressors could produce an excess of glucocorticoid secretion that may act to suppress neurogenesis and decrease dendritic arborization within the hippocampus.

The researchers designed the study as best they could, clearly discussing its limitations, including the small sample sized used. In their conclusion, they clearly state their main findings, pointing out that further confirmatory research is needed. You can ignore the statistics for now, but this journal article is definitely worth reading as an example of a well-conducted study.

2.6 CONCLUSION

In this chapter, we went beyond the news headlines and related articles and learned how to think critically about the quality of research presented in scientific journals. Just because a study appears in a scientific journal doesn't mean its results and/or interpretation of these results should be accepted uncritically. By going to the source of the research, we removed our reliance on the media to assess the quality of the research for us. In the process, we developed our critical thinking skills and learned some basic—but very important—concepts in statistical thinking.

The aim of a good researcher should be to minimize all the possible errors or bias in the study design that could affect the quality of measurements collected. Later in the book, we will question whether other important assumptions about the quality of the data were considered when making statistical decisions. In our next chapter, we will think about what is important in determining the quality of a poll or survey.

2.7 REAL-WORLD EXERCISES

1. Find a news article that discusses a research topic of interest to you. Find the name of the main researcher and the journal in which the research was published. Go to your school library online and search for the related journal article. Search for the journal, and then search for the name of the researcher within the journal.

    a. Critique and question the quality of the research by reading the journal article.

    b. After reading the journal article, question how well you think the news article reported the results of the study.

2. Find two news articles from two different news sources reporting on the same piece of research. Find the articles, and read the related journal articles. Discuss any differences between the two news sources on how well they reported on the piece of research.

3. Design a randomized experiment on a topic of interest to you. Discuss the following:
    a. Who were the subjects, and how were they selected?
    b. What were the treatments the subjects were assigned to?
    c. How were the subjects randomly assigned to treatments?
    d. Did the treatments include a placebo or control group and why?
    e. What was the measurement used, and how was it measured?
    f. Was the experiment single-blinded, double-blinded, or not blinded?
    g. What were the explanatory and the response variables?
    h. What population can the results be extended to?

4. Design an observational study on a topic of interest to you. Discuss the following:
    a. Who were the subjects, and how were they selected?
    b. What were the explanatory and the response variables you were observing?
    c. How did you measure what you wanted to measure?
    d. What were the potential confounding variables you accounted for?
    e. What population can the results be extended to?

5. Find a journal article based on a randomized experiment of interest to you. Discuss the following:
    a. What was the objective of the research?
    b. What were the treatments subjects were randomly assigned to?
    c. Did the researchers check whether the randomization process worked?
    d. Did the researchers include a placebo or control group?
    e. What was the response variable?
    f. How were the measurements taken?
    g. What population did the researchers extend the results of the research to?

6. Find a journal article based on an observational study of interest to you. Discuss the following:

    a. What was the objective of the research?

    b. How were the subjects selected?

    c. What was response variable?

    d. How were the measurements taken?

    e. Did the researchers control for any potential confounding variables?

    f. What population did the researchers extend the results of the research to?

CHAPTER 3

ASSESSING THE QUALITY OF POLLS AND SURVEYS

The purely random sample is the only kind that can be examined with confidence by means of statistical theory, but there is one thing wrong with it. It is so difficult and expensive to obtain for many uses that sheer cost eliminates it.

—Darrell Huff, How to Lie with Statistics

3.1 INTRODUCTION

A **survey** is a type of research, a collection of (focus/related) questions asked of **respondents**. Polls are just a type of survey, a term used when asking an electorate which candidate they will vote for or how they are going to vote in a referendum.

As with any research study, **polling** and survey accuracy depends upon the quality of the sample selected. We want a sample that is truly representative of the population of interest. This means that the sample should be a reflection of the population broken down by gender, age group, location, political views, religious affiliation, and so on. A random sample, where every individual in the population has an equal chance or probability of being selected, should provide the representative sample we require.

The first pollster to truly understand the power of random sampling was George Gallup. He made his reputation when he correctly predicted the outcome of the 1936 presidential election between Franklin Roosevelt and Alf Landon. The *Literary Digest*, a respected magazine at the time, polled around 2.3 million people, predicting

KEY TERMS

Polling: The term used when we ask a sample of respondents a simple question like "Who are you going to vote for in an election?"

Survey: The term used when we ask a sample of respondents a wide range of questions

95% Confidence: A level of confidence as to whether the population parameter of interest is within a range or interval of plausible values

Margin of Error: Measures the accuracy of sample statistics. It is used to construct an interval of plausible values for the population parameter of interest

Respondent: An individual who answers the poll/survey questions

Response Rate: The percentage of individuals in the sample who answer the poll/survey questions

Weighting: A technique used to adjust the results when the sample of respondents is not representative of the population

Online Poll/Survey: The respondents volunteer (self-selected) to take the poll/survey, resulting in a convenience sample that is not representative of the population

Telephone Poll/Survey: A sample of individuals are selected and contacted by telephone

Stratified Sampling: The population is first broken up into groups, and then a random sample is selected from each group

Oversampling: The researcher collects a larger (than representative) proportion of a subgroup of a population. This is to ensure that the margin of error is not too large when analyzing the results of the subgroup

Confidential Poll/Survey: The researcher collects identifying information about the respondent but agrees not to share it

Anonymous Poll/Survey: The researcher does not collect identifying information about the respondent

Leading Question: A question that may lead the respondent to answer the question in a particular way

Double-Barreled Question: A question that asks the respondent's opinion on more than one issue but allows for only one answer

---

a landslide win for Alf Landon. Gallup selected a random sample of 3,000 people and predicted Franklin Roosevelt would win. As history tells us, Gallup got it right.

The *Literary Digest* sent its poll to 10 million of its subscribers: a group of individuals not very representative of the general population at the time, since the wealthy were more likely to be magazine subscribers. However, possibly the more influential issue with the poll was that only 2.3 million subscribers responded, a **response rate** of only 23%. Landon supporters were more passionate, wanting change, and thus were more likely to respond to the poll and have their opinion heard. The result was a much higher proportion of Landon supporters among respondents than in the general population. This skewed the sample, giving their opinions greater weight. This made the poll untrustworthy or biased—i.e., you couldn't use it to make accurate predictions about the election outcome. Which is why *Literary Digest* got it wrong.

Now we need to talk about sample size. The key to a good poll or survey is not having a huge sample size, but having a *quality* sample of sufficient size. The closer to random, the better the quality of the sample. In fact, we can say with **95% confidence** that a truly random sample of 1,024 individuals will get us within 3% (known as the **margin of error**) of the true opinion of all individuals in the population. For a sample of 1,024 individuals, the margin of error is calculated as follows:

$$ME = 1/\sqrt{n}$$

Margin of error = 1/square root of sample size
= 1/square root of 1,024
= 0.03 (or 3%)

This calculation is a somewhat conservative, but valid, estimate of the margin of error. Later in the book, we will discuss in depth what we mean when we say with 95% confidence.

In this day and age, it has become increasingly difficult to obtain a truly random sample of individuals. Until the advent of cell phones, samples were selected through the use of random digit-dialing of landline telephone numbers. However, the decline in the use of landlines and in the willingness of individuals to respond has made polling much more difficult. The article from the *New York Times* titled "What's the Matter With Polling", discusses in depth why these issues have become problematic for polling.

As the article points out the rise in the use of cellphones and the low response rates have caused a real crisis in polling. The article goes on to discuss the impact low response rates have on the cost of doing surveys with interviewers having to make a large number of calls to obtain the desired sample size of respondents.

In this chapter, we will discuss the problems related to having a low response rate and what researchers can do to try and resolve these problems. When response rates for a poll or survey are low, the sample of respondents needs to be treated like a convenience sample. We will reason with the quality of a poll or survey by asking the following questions:

- Was a random or convenience sample used?
- If the sample was random, what was the response rate?
- Was **weighting** applied to adjust for known differences between the sample of respondents and the population?
- What are the possible problems with how the researcher asked the questions?

> **WHAT'S THE MATTER WITH POLLING?**
>
> Since cellphones generally have separate exchanges from landlines, statisticians have solved the problem of finding them for our samples by using what we call "dual sampling frames"—separate random samples of cell and landline exchanges. The problem is that the 1991 Telephone Consumer Protection Act has been interpreted by the Federal Communications Commission to prohibit the calling of cellphones through automatic dialers, in which calls are passed to live interviewers only after a person picks up the phone. To complete a 1,000-person survey, it's not unusual to have to dial more than 20,000 random numbers, most of which do not go to actual working telephone numbers. ...
>
> The second unsettling trend is the rapidly declining response rate. When I first started doing telephone surveys in New Jersey in the late 1970s, we considered an 80 percent response rate acceptable, and even then we worried if the 20 percent we missed were different in attitudes and behaviors than the 80 percent we got. Enter answering machines and other technologies. By 1997, Pew's response rate was 36 percent, and the decline has accelerated. By 2014 the response rate had fallen to 8 percent.
>
>
>
> Web Link: http://www.nytimes.com/2015/06/21/opinion/sunday/whats-the-matter-with-polling.html
>
> Search Term: What's the Matter with Polling?

3.2 POLLING THE BREXIT REFERENDUM

One of the most controversial and impacting decisions made by a nation-state in modern times was Britain's decision in June 2016 to leave the European Union (EU). The uncertainty resulting from the decision has had rippling effects throughout the European economy and beyond. The country will be wrestling with the implications from this decision for a long time to come. How well did the polling do at predicting the outcome leading up to the referendum?

In 2015, the polls did a very poor job of predicting the outcome of the UK general election, so there was great interest in seeing how well they did at predicting the Brexit referendum. The *Financial Times* of London put together a full list of the results of all polls completed prior to the referendum on June 23, 2016. A partial list is shown in Table 3.1.

Table 3.1: Partial List of Individual Brexit Polls

Remain %	Leave %	Undecided %	Date	Pollster	Sample
55	45	0	Jun 22, 2016	Populus	4,700
48	42	11	Jun 22, 2016	ComRes	1,032
41	43	11	Jun 22, 2016	TNS	2,320
44	45	11	Jun 22, 2016	Opinium	3,000
51	49	0	Jun 22, 2016	YouGov	3,766
52	48	0	Jun 22, 2016	Ipsos MORI	–
45	44	11	Jun 20, 2016	Survation	1,003
42	44	13	Jun 19, 2016	YouGov	1,652
53	46	2	Jun 19, 2016	ORB	800
45	42	13	Jun 18, 2016	Survation	1,004
44	44	12	Jun 17, 2016	Opinium	2,006
42	44	14	Jun 17, 2016	YouGov	1,694
44	43	13	Jun 16, 2016	YouGov	1,734
46	43	11	Jun 15, 2016	BMG Research	1,064
42	45	13	Jun 15, 2016	Survation	1,104
41	51	9	Jun 15, 2016	BMG Research	1,468

To see the full list, please scan QR code, or click on the link from a digital copy of the book.

The result of the referendum was 48% voted to Remain part of the EU and 52% voted to Leave. There were 15 polls conducted in the last week before the referendum, 10 predicting Remain and 5 predicting Leave. The very last poll, an **online poll** conducted by a research company called Populus, was the most erroneous, predicting a 10% advantage for Remain over Leave (55% versus 45%).

So why did Populus get it so wrong, especially when the results were based on one of the largest sample sizes (4,700 individuals) of any poll completed before the referendum?

The main problem is that an online poll is based on a convenience sample. The participants were self-selected and therefore not representative of the overall population. Consequently, Populus had to apply a good deal of weighting to the sample, making assumptions about the electorate that were wrong. Populus tried to provide an answer to what went wrong, responding quickly with a detailed explanation.

BREXIT POLL TRACKER

Full list: https://ig.ft.com/sites/brexit-polling/

Search Term: Brexit Poll Tracker

Final Referendum Poll

Since the referendum we have, of course, intensively analyzed why our poll was so wrong. We have peeled back the data, layer by layer, to look at the effect of each of the different methodological stages and the weightings that we applied to the raw data.

We conclude that some of our innovations were right. In particular, ensuring that poll samples define their national identity ('English/Scottish/Welsh only', 'English/Scottish/Welsh more than British', 'British only' etc.) in the same proportions as the population proved important. For polling on political issues, online samples tend to include too many respondents who align their national identity with England (or Wales) rather than Britain—and these people are disproportionately hostile to the EU. The polling organization that ended up closest to the final result introduced this important change to its methods during the referendum campaign, informed by Populus analysis published in March ('Polls Apart').

But two methodological steps were wrong—and caused us to overstate support for remaining in the EU by over 6%. These were the ways that we tried to take account of how undecided voters would end up voting, and the way that we estimated the likelihood to vote of different groups.

FINAL REFERENDUM POLL—POPULUS

Web Link: https://www.populus.co.uk/insights/2016/06/final-referendum-poll/

Search Term: Final Referendum Poll, Populus

Populus goes on to describe the two mistakes they made in their methodology. First, Populus asked the "don't know" responders three additional questions. Analysis had shown that the answers to these questions were strong predictors of how they would vote. This method did not turn out to work very well, putting far too many "don't know" responders in the Remain column.

Second, Populus assumed the proportion of people who would turn out to vote would be the same as in the 2015 general election. However, they also assumed that any increase in turnout "would occur fairly evenly across all demographic groups. The turnout weighting had the effect of further increasing the apparent Remain vote share in our poll—and, therefore, making our poll more wrong still."

This is a good opportunity to discuss the concept of weighting. Weighting is a technique used in the analysis of both online and telephone polling. In an online poll, it is used to adjust the results for the fact that the individuals taking the poll were self-selected, and will not be representative of the population. In a **telephone poll**, it is used to adjust for a low response rate. We will use a simple hypothetical example of a telephone poll to explain the concept.

Say we select a random sample of 1,000 individuals for a telephone poll from a population that is known to contain 50% males and 50% females. If the sample is truly random, it should consist of approximately 50% males and 50% females. In other words, the sample will be an accurate reflection of the percentage of males and females in the population. Let's say only 600 of the 1,000 individuals respond to the survey, a response rate of 600/1000, or 60%. Of the 600 responders, 150 individuals are male, or 25% of the responders. The other 450 individuals are female, or 75% of the responders.

Table 3.2: Percentage of Males and Females in Population and Sample of Responders

Gender	Male	Female
Population	50%	50%
Sample (Responders Only)	25%	75%

To adjust for the underrepresentation of males in our sample, male responses are given a weight of 50/25, or 2. This means that each male response is treated like two male responses in the statistical analysis of the data.

To adjust for the overrepresentation of females in our sample, female responses are given a weight of 50/75, or 0.66. This means that each female response is treated like two-thirds of a female response in the statistical analysis of the data.

In the Populus online poll, they were very proud of the fact that they weighted the responses to align with national identity (English, Scottish, Welsh) in the British population. However, making this adjustment does not mean that they were necessarily going to get a more accurate result. This was an online poll, and these types of polls can result in samples that are far from random and therefore not representative of the population. An online poll is a convenience sample and needs to be weighted by demographic characteristics other than national identity.

For example, in England, the majority of younger voters voted to Remain and the majority of older voters voted to Leave. If younger voters were overrepresented in the online poll, then this would have biased the results toward a Remain result. Populus did state that they adjusted for demographic characteristics, which may or may not have included age groups.

Weighting based on known factors about a population can certainly result in more accurate predictions from an online poll, but it has its limitations. For example, what if there was a particular factor about young people that affected how they would vote in the Brexit referendum that the pollsters didn't know about. Maybe young people who attended college were more likely to vote to Remain. If they were overrepresented in the sample, and the pollsters don't adjust for this factor, then the results of the poll would be biased in favor of Remain. It is possible that there are factors that the pollsters didn't or couldn't adjust for, that affected how particular people voted.

Only a truly random sample is representative of the overall population, resulting in no need for weighting. The farther from random a sample is, the more weighting necessary. For the Populus poll, some of their weighting assumptions were appropriate, but other assumptions (as they admitted) were problematic, leading to a poor prediction of the outcome of the referendum.

A *New York Times* article titled "We Gave Four Pollsters the Same Raw Data. They had Four Different Results." discusses how four pollsters analyzing the same 2016 US Election polling data came up with four different results. Each pollster

WE GAVE FOUR GOOD POLLSTERS THE SAME RAW DATA. THEY HAD FOUR DIFFERENT RESULTS

Well, well, well. Look at that. A net five-point difference between the five measures, including our own, even though all are based on identical data. Remember: There are no sampling differences in this exercise. Everyone is coming up with a number based on the same interviews.

Their answers shouldn't be interpreted as an indication of what they would have found if they had conducted their own survey. They all would have designed the survey at least a little differently — some almost entirely differently.

But their answers illustrate just a few of the different ways that pollsters can handle the same data — and how those choices can affect the result.

So what's going on? The pollsters made different decisions in adjusting the sample and identifying likely voters. The result was four different electorates, and four different results.

Web Link: http://www.nytimes.com/interactive/2016/09/20/upshot/the-error-the-polling-world-rarely-talks-about.html

Search Term: We Gave Four Good Pollsters the Same Raw Data

FLASHPOINTS IN POLLING

Can polls be trusted? This question is on the minds of seemingly everyone who follows the 2016 campaign, though it is hardly unique to this election cycle. The answer is complicated, thanks to myriad challenges facing polling and the fact that pollsters have reacted to these challenges in disparate ways.

Some polls are conducted literally overnight with convenience samples and undergo little or no adjustment. Others are painstakingly fielded for days or even weeks with robust designs and may be adjusted using cutting-edge techniques. These dramatic differences, which have been shown to affect accuracy, are often opaque to news consumers. What follows is a big-picture review of the state of polling, organized around a number of key flashpoints with links to references and research for those who want to better understand the field.

Web Link: http://www.pewresearch.org/2016/08/01/flashpoints-in-polling/

Search Term: Flashpoints in Polling

will make their own decision with regard to how the data should be weighted. As the article points out, these decisions can make a difference in the reported results.

Polling is an important way to get a sense of what the entire electorate are thinking about a candidate for an election or referendum. As already stated, the challenges of polling have definitely increased in recent years. The *Pew Research Center* put together a review of the state of polling; called "Flashpoints in Polling" that is definitely worth reading.

The review discusses the issues with polling: the decline in landlines, the rise of cell phones, low response rates, and others. It also discusses research into trying to resolve these issues.

3.3 DON'T ASK, DON'T TELL (DADT) SURVEY

We will now question the results of a survey related to an issue that was very controversial at the time. In June 2010, the results of a military survey were released. The aim of the survey was to gauge the effects on the military of repealing the "Don't Ask, Don't Tell" policy. The results of the study were discussed in the *New York Times* article titled "Pentagon Sees Little Risk in Allowing Gay Men and Women Serve Openly".

The main issues with this survey were the enormous sample size and the low response rate. The *New York Times* article states that the survey was based on a sample of 115,000 active-duty and reserve service members. Our intuition tells us that the results of the research must be accurate since they were based on such a large sample size. However, quantity does not necessarily mean quality. A large sample that is far from representative of the overall population can yield poor-quality results. We will go

---

PENTAGON SEES LITTLE RISK IN ALLOWING GAY MEN AND WOMEN TO SERVE OPENLY

In an exhaustive nine-month study on the effects of repealing "don't ask, don't tell," the 17-year-old policy that requires gay service members to keep their sexual orientation secret or face discharge, the authors concluded that repeal would in the short run most likely bring about "some limited and isolated disruption to unit cohesion and retention." But they said those effects could be mitigated by effective leadership. ...

In a survey of 115,000 active-duty and reserve service members, the report found distinct differences among the branches of the military, particularly in the Marine Corps, whose leaders have been the most publicly opposed to allowing gay and bisexual men and women to serve openly. While 30 percent of those surveyed over all predicted that repeal would have some negative effects, 40 percent to 60 percent of the Marine Corps and those in various combat specialties said it would be negative.

Web Link: http://www.nytimes.com/2010/12/01/us/politics/01military.html

Search Term: Pentagon Sees Little Risk

SUPPORT TO THE DOD COMPREHENSIVE REVIEW WORKING GROUP ANALYZING THE IMPACT OF REPEALING "DON'T ASK, DON'T TELL" VOLUME 1: FINDINGS FROM THE SURVEYS

Westat has conducted surveys of Service members and their spouses designed to measure perceptions of how a repeal of "Don't Ask, Don't Tell" (DADT) might affect military readiness, military effectiveness, unit cohesion, morale, family readiness, military community life, recruitment, and retention. The surveys were not designed to be a referendum on the issue of DADT repeal, nor can survey results alone answer the question of whether repeal should or should not occur. The surveys can, however, contribute to the decision-making process by providing information on what Service members and their spouses think will be the likely impact of repeal.

Web Link: http://archive.defense.gov/home/features/2010/0610_dadt/Vol%201%20Findings%20From%20the%20Surveys.pdf

Search Term: Support to the DoD Comprehensive Review—Volume 1

beyond the news headline to the source of the research to get a better understanding of the quality of the research and the sample used. The source of the research can be found on the Department of Defense's website.

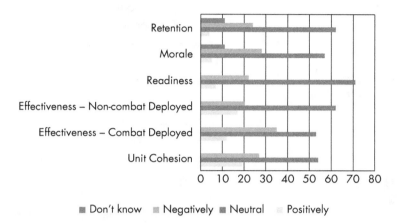

Figure 3.1: Expected Impact of DADT Repeal Across the Main Subject Areas

Source: http://archive.defense.gov/home/features/2010/0610_dadt/

Figure 3.1 shows, according to the results of the survey, the expected impact of repealing DADT. For example, 19% of respondents said repeal would have a positive impact on unit cohesion; 27% said it would have a negative impact; and 54% said it would have a neutral impact.

The survey sample consisted of both Active Duty Service members, and Reserve and National Guard members. Table 3.3 shows a breakdown of population and sample sizes for both groups:

Table 3.3: DADT Survey—Population and Sample Sizes

	Active Duty Service	**Reserve and National Guard**
Population Size	1,416,741	831,193
Sample Size	199,962	199,894

The overall response rate was around 28.7% (of the total sample size of 399,856), resulting in a sample of approximately 115,000 active-duty and reserve members, as stated in the news article. We will discuss why the low response rate is a problem. However, the first question we should ask is why such a large sample of individuals was selected. We discussed in the introduction to this chapter that a properly selected random sample of 1,024 participants will result in a margin of error of 3%. So why were the sample sizes so large in this survey?

DOD COMPREHENSIVE REVIEW—APPENDIX

This tool is based on the multivariate allocation algorithm described by Chromy (1987). This allocation method identifies the smallest total sample size that can be allocated to each stratum so that the margin of error does not exceed precision constraints. The use of the Sample Design Tool produced sample sizes satisfying the constraints that the expected maximum margins of error for proportions estimated for the identified domains of interest were less than or equal to 5%. ***These sample sizes were then increased at the request of the CRWG***. Overall, the final Service member survey sample included 399,856 Active Duty and Reserve Component Service members.

Web Link: http://archive.defense.gov/home/features/2010/0610_dadt/

Search Term: Support to the DoD Comprehensive Review—Appendix

In Appendix A–Survey Methods, a partial explanation is given. The Comprehensive Review Working Group (CRWG), appointed by the secretary of defense, worked with the Defense Manpower Data Center (DMDC) to determine the necessary sample sizes. The CRWG wanted to use **stratified sampling**. This means that the population of service members was split into groups before selecting a sample from each group. The groups were based on the many variables listed in Table A.2 presented in the appendix.

For the Active Duty Services members, the analysis was broken down by Service: 5 levels; Pay Grade: 5 levels; Department of Defense Duty Occupation Group: 2 levels; Location: 2 levels; and Family Status: 2 levels—a total of 16 levels.

For the Reserve and National Guard members, the analysis was broken down by Reserve Component: 7 levels; Reserve program: 3 levels; Pay Grade: 5 levels; and DoD Duty Occupation Group: 2 levels—a total of 17 levels.

There is a good reason for the CRWG wanting to break the analysis down by numerous subgroups. It enables them to obtain sample estimates of opinions on DADT for subgroups of the population along with sample estimates for the entire population. As stated in Appendix A of the report, the DMDC used a sampling tool specifically designed for the DMDC by the Research Triangle Institute.

The numerous subgroups would certainly have resulted in a larger sample size than what would be necessary for analyzing the overall group. We are told that the CRWG requested that the sample sizes be increased above the calculated sizes necessary to maintain the required margin of error of 5% stated in the appendix. But we are not told by how much they were increased, and we are not given a reason why the CRWG made this request.

We can calculate a rough estimate of what the total sample size may have been before the CRWG requested the increase. Using the formula discussed in the introduction to this chapter, a random sample of 400 people from any population will result in the (required) margin of error of 5%. For the Active Duty Service members, there were a total of 16 levels within the 5 subgroups, leading to a required sample size of 6,400 members. For the Reserve and National Guard members, there were a total of 17 levels within the 4 subgroups, leading to a required sample size of 6,800 members. The report mentions that some **oversampling** was done.

Even allowing for oversampling, weighting, and analyzing the data across all subgroups, a conservative estimate of the required sample sizes could be no more than 10,000 for both the Active Duty Service members and the Reserve and National Guard. It would be very interesting to know what the (original) calculated sample sizes were, and why the CRWG asked the DMDC to increase them.

The very large sample size made it much more difficult to complete a quality survey. The survey for Active Duty Service members was initiated through a secure website on July 7, 2010, and five e-mail reminders were sent between July 19 and August 12, 2010. Also, for the survey, independent communications were sent to encourage participation.

As mentioned previously, the overall response rate was 28.7%. If the DMDC had used the samples sizes they calculated originally and the Active Duty Service members were given more time to complete the survey, then the response rate could have been much higher.

Due to the low response rate, the analysis was weighted by a number of demographic characteristics, including age group, gender, race, and pay grade. However, with such a controversial and emotional issue, there were surely other factors that affected how individuals responded to the survey. If any group of individuals (based on these other factors) was under- or overrepresented in the sample, the results would be biased.

Another issue with the survey is that nowhere in the report does it mention whether the survey was **confidential** or **anonymous**. A confidential survey means that the researcher knows the respondent's identity, but it is kept private. An anonymous survey means that the researcher does not know the identity of the respondent.

The service members in the sample were contacted directly through e-mail, suggesting that the survey was not anonymous. As already discussed, the results were stratified by the service level, pay grade, etc. It would be impossible to stratify without knowing who was taking part in the survey. The information could be collected in an anonymous survey, but this could lead to poor-quality data, since there would be no way of knowing whether the responses were valid.

The purpose of the survey was to try and assess whether the army was ready to remove the anonymity of sexual orientation from their hiring process. With such a highly sensitive topic, it would have been very important for the service members to feel comfortable expressing their opinion. If the survey was neither anonymous nor confidential, it could result in many service members choosing the Neutral option to the questions asked. The results of the study (see Figure 3.1) show that there was a very high proportion of individuals choosing Neutral. Maybe the service members who responded to the questions by choosing Neutral felt that the removal of the DADT policy would make no difference one way or the other. Or maybe they made the safe choice. If their responses were confidential, this would have removed these sorts of questions regarding why they chose the Neutral option or any other option for that matter.

There is one group of individuals that would probably have strong opinions on DADT and were most likely underrepresented in the sample. The survey was completed through a secure website, making it more difficult for service members in the field of combat to complete the survey. It is hard to complete an online survey with limited internet access. Therefore, it is quite likely that this specific group of service members was underrepresented in the survey. As already stated, the analysis was weighted by age, gender, race and pay grade to adjust for the low response rate. However, if we take two active service members (of the same age, gender, race, and pay grade), one sitting at a desk in North Carolina, the other facing possible death on the battlefield, there is a good chance that their opinions about DADT could be quite different. It is fair to assume when faced with the possibility of dying on the battlefield that opinions on sexual orientation would fade away very quickly.

Finally, there is no mention in the report or the appendix as to whether or not the samples selected were random samples. As we have discussed previously, it can be very difficult for a researcher to select a random sample from a population because the researcher usually does not have access to the entire population. In the DADT study, one would assume that the entire population of service members was known and in their database. Selecting a random sample—or at least one close to random—should have been possible.

The purpose of collecting a sample of data is to estimate truth in populations. No matter what your opinion regarding sexual orientation or your political affiliation, the only truth that should have mattered in this survey was whether the service members were ready for the removal of the DADT policy. A significant proportion of service members are from families with strong religious backgrounds where certain beliefs regarding sexual orientation would be ingrained since childhood. You can't ask people to simply let go of their beliefs because the majority of the electorate have come to see this issue from a different point of view. Forcing a service member to accept a point of view they were not ready to accept could have led to problems with unit cohesion and morale. The truth, whatever it may be, should be more important than making sure you get the study result that you want.

3.4 POLITICAL POLARIZATION AND MEDIA HABITS

With the outcome of the US presidential election in 2016, there is little doubt that the country is divided. A survey conducted in 2014 by the *Pew Research Center* called "Political Polarization and Media Habits" provides an interesting window into the role our media habits may have played in causing this division.

The survey was based on a total sample size of 2,901 respondents out of a total of 4,753 invited to take part, a response rate of 61%. The margin of error was 2.3%. The respondents were recruited from a larger group of 10,013 Americans referred to as the American Trends Panel, a randomly selected, representative sample of Americans. Given that particular smaller subsets of groups (from the larger group) were more likely to volunteer for the survey, a multistep weighting process was applied to adjust for that fact. This weighting process was described in detail in the final report. Overall, this is a good example of a well-conducted survey.

The survey responders were placed in five ideological groups (consistent liberals, mostly liberals, mixed, mostly conservatives, and consistent conservatives) based on questions on a range of political values. The researchers completed an extensive survey asking the respondents questions like the following:

1. What is your main source of news?
2. Which of the following list of 36 news sources have you heard of?
3. What is your level of trust in each of the news sources you have heard of?
4. How many posts do you see on social media that are related to politics?
5. Do you share, like or comment on the posts that you see on social media?

The questions resulted in many interesting statistics:

1. 47% of consistent conservatives cited Fox News as their main source of news, a percentage far greater than any other news source. Consistent liberals relied on a larger range of news sources.

2. The most trusted news sources for each of the groups were as follows: 72% of consistent liberals trusted NPR; 66% of mostly liberals trusted CNN; 61% of mixed trusted CNN; 72% of mostly conservative trusted Fox News; and 88% of consistently conservative trusted Fox News.

3. Consistent conservatives saw more Facebook posts in line with their views than consistent liberals, 47% versus 32%, respectively.

4. Consistent liberals more likely to block others because of politics than consistent conservatives, 44% versus 32%, respectively.

In Chapters 7 and 8, we will analyze some of these percentages (the sample statistics) to make

POLITICAL POLARIZATION & MEDIA HABITS

The project—part of a year-long effort to shed light on political polarization in America—looks at the ways people get information about government and politics in three different settings: the news media, social media, and the way people talk about politics with friends and family. In all three areas, the study finds that those with the most consistent ideological views on the left and right have information streams that are distinct from those of individuals with more mixed political views—and very distinct from each other.

WebLink: https://www.journalism.org/2014/10/21/political-polarization-media-habits/political-polarization-and-media-habits-final-report-7-27-15/

Search Term: Political Polarization & Media Habits Final Report

conclusions about the possible values of these percentages in the population (the population parameters).

There are many other interesting findings to be found in the report. One of particular interest is the graphical presentation of the level of trust each of the five groups of respondents have in the thirty-six different news sources. The one and only news source trusted more than distrusted by all five groups was the *Wall Street Journal*. Google News, a news aggregator, was a close second, with only consistent conservatives saying they trusted it just as much as they distrusted it. This is a strange finding seeing Google News is simply a news aggregator. BuzzFeed was the only news source more distrusted than trusted by all five groups. However, the researchers do point out that only about a third of all respondents had heard of BuzzFeed, so these strong views are based on a relatively small sample size.

As with any effect that we observe in the world around us, the causes of the outcome of the US presidential election are complex, with many factors involved. However, there is little doubt that our current media habits are a major factor contributing to political polarization in the United States. The results of these types of surveys can help us to better understand the role that media consumption plays in this division and perhaps be used as a starting point to come up with insights or ideas that may help heal that divide.

The media has a powerful influence over our lives because we have to rely on it for facts and information about what is happening in the world. It is very important to learn how to think for ourselves in order to see through the noise and get at the truth. However, we need all individuals working in the media to return to having some semblance of integrity by simply pursuing the objective truth and reporting all the facts. If we have all the facts, then we can discuss and make our own conclusions.

3.5 ASKING QUESTIONS

HOW ARE POLLS CONDUCTED?

Web Link: http://media.gallup.com/PDF/FAQ/HowArePolls.pdf

Search Term: Gallup How Are Polls Conducted

In the DADT survey, we discussed how the active-duty and reserve service members' response to the questions may have been affected by apparent lack of confidentiality or anonymity of the survey. There are other ways in which the questions asked could lead to an answer that is not the true opinion of the respondent.

For example, in an election poll, we want the respondent to make the same choice of candidate they will make on Election Day. The polling organization Gallup uses the following rule:

Gallup's rule in this situation is to ask the question in a way that mimics the voting experience as much as possible. We read the names of the presidential and vice presidential candidates, and mention the name of the party line on which they are running. All of this is information the voter would normally see when reading the ballot in the voting booth.

They are many good resources for creating surveys that point out the possible pitfalls in the designing of your questions. One research company named Qualtrics provides tips for writing survey questions in an article titled "Survey Questions 101: Do You Make Any of These 7 Question Writing Mistakes?"

Some of the mistakes Qualtrics points out are the use of **leading questions** such as "How would you rate the career of legendary outfielder Joe DiMaggio?" or **double-barreled questions** such as "How likely are you to go out for dinner and a movie this weekend?"

Avoid use of the word *and*, which results in asking two questions in one. Avoid questions that contain statements like *not unnecessary* or *wasn't irresponsible* that can be extremely confusing. Avoid questions that include statements like *would you be willing* or *do you agree* that may elicit a desire to please on the part of the respondent. Finally, try to imagine standing in the shoes of the respondent and hearing the question from their point of view.

> **SURVEY QUESTIONS 101: DO YOU MAKE ANY OF THESE 7 QUESTION WRITING MISTAKES?**
>
> Details, details, details. Creating surveys that yield actionable insights is about details. And writing effective questions is the first step. We see common mistakes that keep survey questions from being effective all the time.
>
>
>
> Web Link: https://www.qualtrics.com/blog/writing-survey-questions/
>
> Search Term: Qualtrics Writing Survey Questions

3.6 CONCLUSION

Deciding on whether a poll or survey is worth paying attention to is about asking the right questions. We want to know the sample size used and the margin of error associated with the sample estimates. For telephone polling or surveys based on a random sample, we want to know the response rate and the sort of weighting applied to account for a low response rate. For online polling and surveys, we want to know what sort of weighting was applied to the data due to the nonrandom or convenient nature of the sample.

Polling and surveys can be a powerful way of getting a sense of how the population is thinking about certain social or political issues. A relatively small sample size can get us a sample estimate quite close to the true opinion of the entire population, no matter what size the population may be. However, if the sample is far from random, the response rate is low and the weighting applied is insufficient or nonexistent, then it is simply a case of "garbage in, garbage out." The results of such a poll or survey may be very far from the true opinion of the population.

Everyone has different opinions on the issues, but we are also different in many other ways. In the next chapter, we will learn how to reason with variation in measurements across individuals.

3.7 REAL-WORLD EXERCISES

1. Find a poll in the media. Two good sources are https://projects.fivethirtyeight.com/polls/ and https://www.realclearpolitics.com/epolls/latest_polls/. Assess the quality of the poll by asking the following questions:

   a. Did the researchers use a random sample or a convenience sample?

   b. What was the population the results were extended to?

   c. What was the sample size and the margin of error?

   d. Was weighting applied? If so, what factors were included in the weighting?

   e. What were the exact questions asked? Discuss any problems with the wording of the questions.

2. Design, execute, and write up a sample survey on a topic that interests you. Design your questionnaire to have at least ten questions. Include at least one hundred people in your survey.

   a. Include questions that result in quantitative (or measurement) data (age, weight, height, salary, etc.) and categorical data (Yes/No, Male/Female, etc.). The categorical data should be both nominal (example: race [American Indian or Alaska Native, Asian, Black or African American, Native Hawaiian or Other Pacific Islander, and White]) and ordinal (example: Strongly Agree, Agree, Neutral, Disagree, Strongly Disagree).

   b. Include at least two questions that result in quantitative variables that you feel may be related. Example: time studying and homework score.

   c. Include at least two questions that result in categorical variables that can be displayed in a two-way table (example: Male/Female or Yes/No).

   d. Include at least two questions that result in quantitative variables for making comparisons between groups (example: compare length of time studying per week for Males and Females).

   e. Discuss the purpose of the study and how the data was collected. Discuss any limitations of the data due to difficulties in data collection (convenience versus a random sample, low response rate, problems with wording of questions).

3. Find a media source that provide the results of election polling. Two good sources are https://projects.fivethirtyeight.com/polls/ and https://www.realclearpolitics.com/epolls/latest_polls/. Choose two polls from different pollsters based on the same topic or question.

    a. Find the sample sizes used, and calculate the margin of error for both polls.
    b. Discuss the reasons why the results are different across the two polls.
    c. Explain in your own words what the results of the two polls mean.
    d. What population can the results be extended to?

4. Find a survey discussed in the media based on a topic of interest to you. One good resource is http://www.pewresearch.org. Answer the following:

    a. How were the subjects selected?
    b. What were the exact questions asked?
    c. Was type of sample was used. If random, what was the response rate?
    d. Did the researchers apply weighting to the sample? If so, in your own words describe the weighting used.
    e. What population were the results extended to?
    f. How large was the sample, and what was the margin of error?

5. Design a minisurvey on a topic of interest with two categorical questions that you feel may be related. Example: Social Issue—Yes/No and Gender:—Male/Female. Using social media, ask 50 people both questions. Complete the following:

    a. What were the questions you asked?
    b. Create a bar chart or pie chart of responses for both questions.
    c. Calculate the percentage response (of Yes/No, for example) for both questions.
    d. Calculate the margin of error.
    e. What population, if any, can the results be extended to?

6. Find survey data online based on a topic of interest to you. One good source is http://www.pewresearch.org/download-datasets/. Complete the following:

   a. What were the two primary categorical questions asked?

   b. Create a bar chart or pie chart of responses for both questions.

   c. Calculate the percentage response (of Yes/No, for example) for both questions.

   d. Calculate the margin of error.

   e. What population were the results extended to?

IMAGE CREDITS

Fig. 3.1: Copyright © 2016 by Financial Times Ltd.
Fig. 3.2: Source: http://archive.defense.gov/home/features/2010/0610_dadt/Vol%201%20Findings%20From%20the%20Surveys.pdf

CHAPTER 4

VISUALIZING AND SUMMARIZING QUANTITATIVE DATA

We must understand variation.

—W. Edwards Deming

4.1 INTRODUCTION

Variation is everywhere. It is in us, around us, and in everything we do. The random processes of evolution over millions of years have resulted in a richly varied world. For example, people differ or vary in countless ways:

> *Race, Gender, Skin Color, Eye Color, Sexual Orientation, Beliefs, Opinions, Weight, Height, Waist Size, Blood Pressure, Heart Rate, Cholesterol Level, Anxiety Level, Foot Size, Hand Size,...*

Learning how to reason with the variation in measurements is the most important part of what statistical thinking is all about. If everyone were the same height, weight, or had the same response to a particular medication, there would be no need for statistics or statistical thinking. If we were to measure one individual's height, weight, or response to a medication, we would know everything there is to know (with regard to each measurement) about the entire population of individuals. This certainty in measurement would make our lives much easier, but what a boring world it would be!

In this chapter, we will learn how to reason with the variation in measurements across individuals. We will create graphical displays of sample measurements and ask the following questions:

KEY TERMS

(Measure of) Central Tendency: A central or typical value for a distribution of measurements

Histogram: A visual representation of the distribution of individual values across the complete range of possible values of the measurement in question

Normal Distribution: A symmetric (bell-shaped) distribution of quantitative (or measurement) data

Skewed Distribution: An asymmetric distribution of measurements with a small percentage of either low or high values of the measurement

Mean: A measure of central tendency most appropriate for symmetric distributions of quantitative (or measurement) data

Median: A measure of central tendency most appropriate for skewed distributions of quantitative (or measurement) data

Standard Deviation: A measure of the spread of values around the mean most appropriate for symmetric distributions of quantitative (or measurement) data

Interquartile Range: A measure of the spread of values around the median most appropriate for skewed distributions of quantitative (or measurement) data

Sample Variance: The average of the squared deviations around the sample mean

Empirical Rule: Under the normal distribution, 68% of measurements are within one standard deviation of the mean; 95% are within two standard deviations of the mean; and 99.7% are within three standard deviations of the mean

Standardized Score (z-score): For data that follows a normal distribution, a value that is a measure of the number of standard deviations (above or below the mean) a particular value lies

Standard Normal Distribution: A normal distribution of measurements with a mean equal to zero and a standard deviation equal to one

- How do we determine what is a typical value of a measurement?
- How far does a particular individual's measurement deviate from a typical value?
- How spread out are the individual measurements around the typical value?
- How do we measure the spread of individual values around a typical value?
- In what range do most individual values of the measurement lie?

We will learn how many measurements in nature are distributed according to what is known as the **normal distribution**. We will learn how to calculate and reason with summary measures of **central tendency** (a typical value) and spread of individual values for describing data that follows the normal distribution. We will also look at alternative measures of central tendency and spread for measurements that follow what are known as **skewed distributions**. Finally, we will learn how to calculate the probability of obtaining an individual value more extreme than a particular value.

4.2 HISTOGRAMS AND THE NORMAL DISTRIBUTION

We will start to reason with the variation in sample measurements by creating a graphical display known as a **histogram**. A histogram is a visual representation of the distribution of individual values across the complete range of possible values of the measurement in question.

Figure 4.1 shows a histogram of men's height (in centimeters) for a random sample of 2,742 individuals. The sample data was collected by the 2011–2012 National Health and Nutrition Examination Survey (NHANES), "a program of studies designed to assess the health and nutritional status of adults and children in the United States. The survey is unique in that it combines interviews and physical examinations."

VISUALIZING AND SUMMARIZING QUANTITATIVE DATA | 59

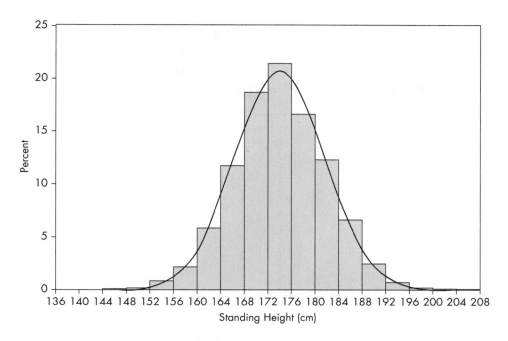

Figure 4.1: 2011–2012 NHANES—Heights of Men in the United States

The horizontal axis represents the data's range of values from smallest to largest. The axis is then broken up into equally spaced intervals, or bins. For the height data, the axis is broken up into intervals of 4 centimeters (cm), the unit of measurement. A bar is drawn above each interval range representing the percentage of individuals in the sample having a height value within that range. The equally spaced intervals ensure that each bar is a fair visual representation of the percentage of individuals with heights within that interval. The taller the bar, the larger the percentage of individuals in the sample with a height value within that interval.

Notice that the distribution of heights has a symmetric bell shape, which peaks at the center of the graph. The interval with the highest percentage of individual heights goes from 172 cm to 176 cm, with approximately 22% of individuals having a height within that interval. Move away from this center bar, and the declining bars show that the percentage of individuals with height values greater than 176 cm or less than 172 cm decreases quite rapidly.

The smooth, symmetric curve drawn on top of the histogram is known as the normal curve, or normal distribution. It is considered an approximate representation of the true population distribution of values based on the sample data selected.

NHANES 2011–2012 EXAMINATION DATA

Web Link: https://wwwn.cdc.gov/Nchs/Nhanes/Search/DataPage.aspx?Component=Examination&CycleBeginYear=2011

Search Term: NHANES 2011–2012 Examination Data

The normal distribution is the most important distribution used in statistics. The distribution of many measurements in nature tends to take on this shape. The histogram of diastolic blood pressures (men and women) from the 2011–2012 NHANES shown in Figure 4.2 below is another example of a measurement whose values are normally distributed.

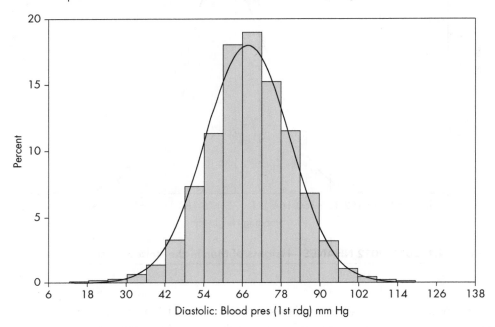

Figure 4.2: 2011–2012 NHANES—Diastolic Blood Pressure

It is the evolutionary nature of how each individual measurement value is determined that produces this result. In the case of diastolic blood pressure, many factors (known and unknown) pull an individual's measurement up and down. The accumulation of all the individual values tends to fall around a central value in the shape of the normal distribution. The fact that the histogram of individual measurements takes on this shape is a very interesting constant to be found in nature. Under the normal distribution, there is an exact relationship known as the **empirical rule** between the **mean** and how far we expect any individual measurement value to be from the mean; measured using what is known as the **standard deviation**.

For data that behaves in this manner, we can summarize the distribution of measurement values with the normal distribution as an approximation for the population distribution. We can summarize our data by calculating measures of central tendency and spread—the mean and the standard deviation, respectively. The mean is considered a typical value for the measurement variable we are looking at. The standard deviation is a measure of how spread out the individual values are around the mean.

4.3 NORMAL DISTRIBUTION: MEASURES OF CENTER AND SPREAD

In Figure 4.1 the individual height values are normally distributed around a central value. Under the normal distribution, the mean is located where approximately 50% of measurements are less than this value, and approximately 50% are greater than this value. The mean is a type of average, a measure of central tendency. The **sample mean** is a summary statistic that is an approximate estimate of the **population mean**.

The sample mean is calculated as follows:

$$\bar{x} = \frac{\sum_{i=1}^{n} x_i}{n}$$

Sample Mean = Sum of Individual Measurement Values/Sample Size

For the NHANES men's height data, the sum of all the height values is 477,277 centimeters (cm), and the sample size is 2,742 men. The sample mean is calculated as follows:

Sample Mean = 477,277/2,742
= 174 cm

In this sample, approximately 50% of men's heights are less than 174 cm, and approximately 50% of men's heights are greater than 174 cm.

While the sample mean is a useful measure for summarizing our data, it tells us nothing about the spread of individual values in our data. For example, say we were told that the sample mean weight loss on a diet program was 20 pounds, based on a sample of 15 individuals. Let's look at (hypothetical) samples from two weight loss programs that resulted in the same sample mean weight loss.

First Weight Loss Program: –15, –10, –5, 0, 5, 10, 15, 20, 25, 30, 35, 40, 45, 50, 55
Second Weight Loss Program: 13, 14, 15, 16, 17, 18, 19, 20, 21, 22, 23, 24, 25, 26, 27

The mean weight loss is 20 pounds for both weight loss programs. However, we can see that there is much greater variation in individual weight losses on the first program than on the second program. On the first program, one individual lost 55 pounds, where another individual gained 15 pounds (negative weight loss of –15 pounds). The individual weight losses on the second program are much more consistent, narrowly spread around the mean. Both programs appear to be equally effective at reducing weight when we simply compare the means. However, it is important that we calculate a statistic that is a measure of the variation in individual weight loss for both programs: in other words, a statistic that measures the spread of individual weight loss values around the mean.

As already stated, the summary statistic used to measure the spread of individual values around the mean is called the standard deviation. It is a measure of the average distance any value is expected to deviate from the mean.

Let's use our NHANES height data to calculate the standard deviation. We start by calculating the individual deviations: each individual height value minus the mean height. For example, an individual with a height of 177 cm will have a deviation of 3 cm (177 minus 174). Another individual with a height value of 172 cm will have a deviation of –2 cm (172 minus 174), and so on.

Since the mean is a measure of central tendency under the symmetric normal distribution, approximately half of the individual deviations are negative numbers, and the other half are positive numbers. The sum of all the individual (positive and negative) deviations adds up to zero, which tells us nothing about the spread of the height values around the mean.

To solve this problem, the individual deviations are first squared and then added together. For example, an individual with a deviation of 3 cm will have a squared deviation of 9 cm squared. An individual with a deviation of –2 cm will also have a squared deviation of 4 cm squared.

The sum of the squared deviations is divided by the sample size minus one, resulting in a statistic called **sample variance**. This calculation can be written as follows:

$$s^2 = \frac{\sum_{i=1}^{n}(x_i - \bar{x})^2}{n-1}$$

Sample Variance = (Sum of Squared Deviations)/(Sample Size – 1)

Variance is an important statistic used in statistical analysis. Later in the book, we will discuss a statistical technique known as analysis of variance or simply ANOVA. It is used for analyzing differences between several means by comparing one measure of variance to another. For our current purposes, variance is simply a necessary step in the calculation of the standard deviation, our measure of variability of measurements around the mean. Note that the naming conventions used in statistics can sometimes be confusing or counterintuitive. Dividing by the sample size minus one instead of the sample size is a mathematical correction that ensures we are calculating a sample variance that is an unbiased estimate of the population variance.

Finally, we take the square root of the sample variance to obtain the sample standard deviation. The fact that we had to square the individual deviations means we have to take a square root to obtain a summary statistic in the units (cm) of the original measurements. For the NHANES height data, the sample standard deviation turns out to be equal to 7.7 cm. Please see the appendix for a more detailed breakdown of how the sample standard deviation was calculated for this example.

We have summarized the distribution of our individual values very concisely. We used a histogram to visualize the distribution of values and used it to approximate the population distribution with the normal distribution. We approximated the population mean with the sample mean of 174 cm and the population standard deviation with the sample standard deviation of 7.7 cm. We will make use of these values in the next section, along with the empirical rule, to further summarize the distribution of measurements under the normal distribution.

In Chapter 2, we discussed a study that looked at the relationship between playing American football and brain injury. The researchers compared players with concussion to players without concussion and compared both these groups to healthy controls with 25 subjects in each group. The key measurement variable was hippocampal volume, an indicator of cognitive ability. For each of the three groups, the sample means and sample standard deviations were as follows:

Table 4.1: Hippocampal Volumes

Group	Sample Mean	Sample Standard Deviation
Controls	7572	1084
Players without History of Concussion	6489	815.4
Players with History of Concussion	5784	609.3

The aim of the study was to examine the effect of football playing on cognitive ability. From the descriptive statistics presented in Table 4.1, we can see that hippocampal volume is lower on average for players without a history of concussion compared to the healthy controls and lower still for players with history of concussion. When we look at our measure of spread of individual hippocampal volumes around the sample means, the sample standard deviation, we see that the spread also decreases across groups. From this initial descriptive analysis of the data, there is definitely indication of an association between football playing and hippocampal volume. Individual hippocampal volumes are falling as we move from controls to players with concussion and are concentrating around a smaller mean value. We will return to this study in a later chapter to discuss the results of the statistical analysis comparing sample means across groups.

4.4 THE EMPIRICAL RULE AND THE STANDARD NORMAL DISTRIBUTION

We will now look at the relationship between the mean and the standard deviation under the normal distribution. The empirical rule for the normal distribution states that:

- 68% of individual values are within one standard deviation of the mean
- 95% of individual values are within two standard deviations of the mean
- 99.7% of individual values are within three standard deviations of the mean

Any individual values that deviate by more than two standard deviations from the mean are considered outlying values (or outliers), since such extreme values would only be observed 5% of the time.

To be more exact, 95% of values are within 1.96 standard deviations of the mean under the normal distribution. For ease of explanation, we round this value to two standard deviations.

We can make use of the empirical rule for the normal distribution to summarize the data further. For the NHANES men's height data (with a sample mean equal to 174 cm and a standard deviation equal to 7.7 cm), we can state that approximately:

- 68% of individual heights are between 166.3 cm and 181.7 cm
- 95% of individual heights are between 158.6 cm and 189.4 cm
- 99.7% of individual heights are between 150.9 cm and 197.1 cm

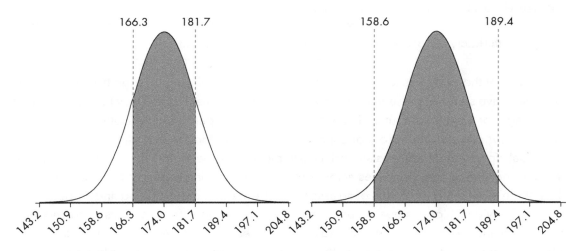

Figure 4.3: The Empirical Rule—2011–2012 NHANES Men's Height

Data source: http://blogs.sas.com/content/sastraining/2014/06/10/producing-normal-density-plots-with-shading/

We can approximate the histogram of heights with the normal distribution. We can see from Figure 4.3 that approximately 68% of individual heights are within one standard deviation of the sample mean height of 174 cm, or between 166.3 cm and 181.7 cm; and approximately 95% of individual heights are within two standard deviations of the sample mean height of 174 cm, or between 158.6 cm and 189.4 cm.

We might also like to know how many standard deviations a particular value is from the mean and what percentage of values are more extreme than any particular value. For our NHANES data, we might want to know how many standard deviations an individual's height is above or below the mean. We might want to know the probability that a person we select from this population will have a height value greater or lesser than a particular height value.

A **z-score (or standardized score)** is a calculation that tells us how many standard deviations an individual value is above or below the mean. A standardized score is a value from the standard normal distribution with a mean equal to 0 and a standard deviation equal to 1. Under the standard normal distribution:

- 68% of height values will have z-scores between –1 and +1
- 95% of height values will have z-scores between –2 and +2
- 99.7% of height values will have z-scores between –3 and +3

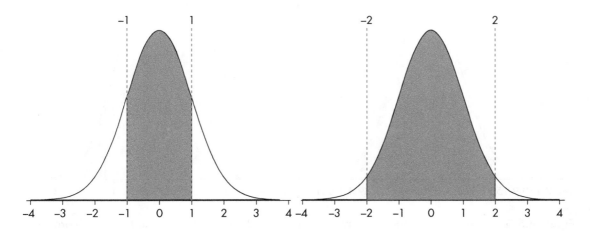

Figure 4.4: The Empirical Rule and the Standard Normal Distribution

Data source: http://blogs.sas.com/content/sastraining/2014/06/10/producing-normal-density-plots-with-shading/

The further an individual measurement value is from the mean, the larger its z-score, positive or negative. The mean of our data will have a z-score of 0 (it is zero standard deviations from itself). The height value of 181.7 cm, one standard deviation above the mean, has a z-score of 1; the height value of 166.3 cm, one standard deviation below the mean, has a z-score of –1, and so on. We can calculate a z-score by subtracting the mean from the particular value and dividing by the standard deviation.

For our NHANES data, a simple question we might ask is how many standard deviations is the height value of 184 cm from the mean height of 174 cm? We can answer this question by calculating a z-score:

$$z = \frac{x_i - \bar{x}}{s}$$

$$\begin{aligned}\text{z-score} &= (\text{value} - \text{mean})/\text{standard deviation} \\ &= (184 - 174)/7.7 \\ &= 1.3\end{aligned}$$

The z-score of 1.3 tells us that the height value of 184 cm is 1.3 standard deviations above the mean height of 174 cm.

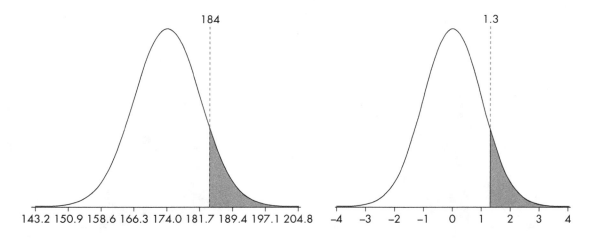

Figure 4.5: 2011–2012 NHANES—Heights of Men in the United States & Standardized Score

Data source: http://blogs.sas.com/content/sastraining/2014/06/10/producing-normal-density-plots-with-shading/

We have learned how to calculate the corresponding z-score for a particular measurement value. We can now use the z-score along with the standard normal distribution, to calculate the chance or probability of an individual having a measurement value more extreme than that particular value. For the NHANES height data, we might want to know what is the probability of an individual having a height greater than 184 cm. We can answer this question by calculating the probability of getting a z-score greater than 1.3 under the standard normal distribution.

Using a standard normal table (which you can find online) or statistical software, we can calculate the probability of getting a z-score of 1.3 or greater. The probability turns out to be equal to 0.10, or 10%. The probability is represented by the gray shaded area in Figure 4.5. This means that (based on our NHANES height data) approximately 10% of individual men's heights are greater than 184 cm. We can use this same approach to calculate the percentage of height values more or less extreme than any particular height in question.

4.5 SKEWED DISTRIBUTIONS: MEASURES OF CENTER AND SPREAD

There are many other types of distributions used in statistics (that go beyond the scope of this book) that are non-normal. The mean and standard deviation are used as measures of central tendency and spread for many of these distributions. However, when the distribution is what is known as a skewed distribution, there are alternative measures of central tendency and spread that we can use.

A skewed distribution is an asymmetric distribution of measurements with a small percentage of either low or high values of the measurement. The more skewed the distribution of measurements are, the less valid the mean and the standard deviation are as measures of center and spread. When there is a small percentage of extreme (low or high) values, the mean gets pulled away from the center of the data, and the standard deviation gets inflated by the small percentage of very large deviations from the mean. As an alternative, we will use the median and the interquartile range as a measure of central tendency and spread, respectively.

The **median** is a measure of the center of our data that is not affected by extreme values. The median is calculated by first ordering the individual values from smallest to largest. If we have an odd number of values in our sample, the median is simply the middle value: the value where fifty percent of ordered measurements are below and fifty percent are above this value. If we have an even number of values in our sample, the median is the sum of the two middle values divided by two.

There is an important distinction between the mean and median as measures of central tendency. The mean is calculated using all the values in our sample. When there is a small percentage of extreme values in our data, the mean is pulled toward those values, away from the center of our data. The median is not affected by extreme values because it is determined by one or two middle values in our data. The median is always located in the center of our data.

68 | STATISTICAL THINKING THROUGH MEDIA EXAMPLES

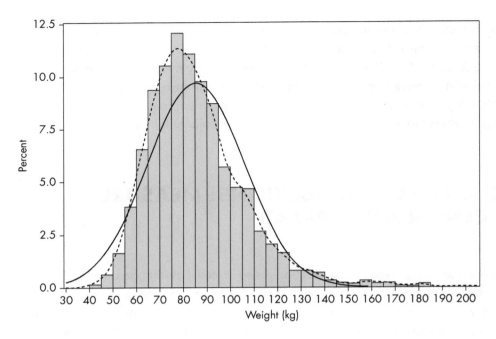

Figure 4.6: 2011–2012 NHANES—Weight of US Men (kg)

Figure 4.6 shows the distribution of 2,742 men's weight in kilos (kg) for the 2011–2012 NHANES survey. The median is equal to 82.3 kg, and the mean is equal to 85.4 kg. The mean is pulled away from the center of the data by the small percentage of extremely heavy men. We can see that the normal distribution (black-lined curve) does not fit the data very well, compared to the height data. The dashed-lined curve tries to approximate the true shape of the population distribution. The weight data is skewed to the right or positively skewed, meaning the data contains a small percentage of large weight measurements. In this case, the mean is not a good measure of the central tendency, and the median should be used in its place. Fifty percent of men's weights are less than the median weight of 82.3 kg, and 50 percent are greater than that value.

The **interquartile range** is a measure of spread of our measurement data that is not affected by extreme values. It measures the range of the middle 50 percent of the data. It is determined by first calculating what are known as the first and third quartiles. When we focus on the lower 50 percent of (ordered) data values, the first quartile is simply the median for these data values. When we focus on the upper 50 percent of (ordered) data values, the third quartile is simply the median for these data values. For the NHANES height data, the first quartile is equal to 71.1 kg, and the third quartile is equal to 95.4 kg.

The value of the first quartile is subtracted from the value of the third quartile, resulting in the interquartile range. The middle 50 percent of weight measurements are between 71.1 kg and 95.4 kg, resulting in an interquartile range of 24.3 kg. The interquartile range, as a measure of spread of the weight measurements around the median, is not affected by the small percentage of extremely heavy men in our sample, since it is based on the middle 50 percent of values.

Natural variation in height is due to genetics and heredity. It is not affected by external factors related to modern-day living. Natural variation in weight is also due to genetics and heredity, but in addition there are also external factors that affect a person's weight, like poor eating habits and lack of exercise.

The *New York Times* article titled "What Eating 40 Teaspoons of Sugar a Day Can Do to You", is about a man who went on a diet that contained the equivalent of 40 teaspoons of sugar a day for 60 days. It is a study with one participant, which does not make it very scientific. However, it is interesting to read how much this change in one factor affected his weight.

Mr. Gameau, interviewed for the article, talks about how he decided to make the film after seeing so much conflicting evidence on the benefits and dangers of sugar. He goes on to describe the effect this one change in his diet had on his body—including signs of acquiring a fatty liver—and how the weight dropped off and all his symptoms went away once he returned to his normal diet.

Another example of a non-normal or skewed distribution is household income. The *New York Times* article titled "U.S. Household Income Rises to Pre-Recession Levels, Prompting Cheers and Questions" discusses the results of a report published by the US Census Bureau titled "Income and Poverty in the United States: 2017." The report was based on the Current Population Survey, a monthly survey of about 125,000 households. As the news article and survey point out, the estimated median household income in the US in 2017 was $61,372, an increase from $60,309 in 2016. However, a closer look at the contents of the report gives a clearer sense of the distribution of household income in the US and how it has changed over the years.

The report includes a table titled "Table A1: Households by Total Money Income, Race, and Hispanic Origin of Householder: 1967 to 2017." The table provides the percent distribution of household incomes (adjusted for inflation) from 1967 to 2017, including means and medians. In the table, we can see that in 2017, 7.7% of households earned an income of $200,000 or greater. This small percentage of big earners results in a distribution of household incomes that is skewed to the right or

---

WHAT EATING 40 TEASPOONS OF SUGAR A DAY CAN DO TO YOU

Soda has been a major target in the debate over sugar and its role in the obesity crisis. But high levels of added sugars can be found in many seemingly healthful foods, from yogurts to energy bars and even whole-grain bread.

……

In "That Sugar Film," which first had its debut in Australia this year, Mr. Gameau gives up his normal diet of fresh foods for two months to see what happens when he shifts to eating a diet containing 40 teaspoons of sugar daily, the amount consumed by the average Australian (and an amount not far from the 28 teaspoons consumed daily by the average American teenager). The twist is that Mr. Gameau avoids soda, ice cream, candy and other obvious sources of sugar. Instead, he consumes foods commonly perceived as "healthy" that are frequently loaded with added sugars, like low-fat yogurt, fruit juice, health bars and cereal.

Web Link: http://well.blogs.nytimes.com/2015/08/14/what-eating-40-teaspoons-of-sugar-a-day-can-do-to-you/

Search Term: What Eating 40 Teaspoons of Sugar a Day Can Do to You

NEWS ARTICLE: U.S. HOUSEHOLD INCOME RISES TO PRE-RECESSION LEVELS, PROMPTING CHEERS AND QUESTIONS

Web Link: https://www.nytimes.com/2018/09/12/us/politics/median-us-household-income-increased-in-2017.html

Search Term: U.S. Household Income Rises to Pre-Recession Levels

NEWS ARTICLE: INCOME AND POVERTY IN THE UNITED STATES: 2017

Web Link: https://www.census.gov/library/publications/2018/demo/p60-263.html

Search Term: Income and Poverty in the United States: 2017

WHAT IS MIDDLE CLASS IN MANHATTAN?

Web Link: http://www.nytimes.com/2013/01/20/realestate/what-is-middle-class-in-manhattan.html?_r=0

Search Term: What Is Middle Class in Manhattan?

towards these higher incomes. The mean household income of $86,220 is well above the median of $61,372, indicating that the small percentage of very large household incomes are pulling the mean far from the center of the distribution of household incomes. In fact, when we look at the difference between mean and median household income from 1967 to 2017, the difference has been consistently increasing over the five decades. In 1967, only 1.1% of households earned an income of $200,000 or greater. At the time, the mean was only $5,444 greater than the median while in 2017, the mean was $24,848 greater than the median. This is a strong indicator that more and more of the wealth in the US is becoming concentrated in the hands of the few.

We will now look at an example from the media where the mean was incorrectly used as a measure of central tendency. According to an article in the *New York Times*:

What Is Middle Class in Manhattan?

The average Manhattan apartment, at $3,973 a month, costs almost $2,800 more than the average rental nationwide. The average sale price of a home in Manhattan last year was $1.46 million, according to a recent Douglas Elliman report, while the average sale price for a new home in the United States was just under $230,000.

When we click the link in the article to the Douglas Elliman report, we see that the average sale price of a Manhattan apartment the article refers to is the mean and not the median. The mean sale price is $1.46 million, but the median sale price is only $837,000. The fact that the mean is much larger than the median indicates that the distribution of sales prices is skewed to the right or positively skewed. There is a small percentage of extremely expensive apartments in Manhattan pulling the mean away from the center of the data.

The most extreme example is the penthouse Manhattan apartment that went for $238 million in 2019, by far the highest price of a home in the country. This apartment price is what we call an extreme outlier, pulling the mean apartment price well above the median apartment price.

The median is unaffected by the small percentage of extremely expensive apartment prices and should be used as the measure of central tendency. Fifty percent of apartment prices in Manhattan are greater than $837,000, and 50% are less than that value. In this case, the median should be considered a typical price for an apartment in Manhattan and not the mean sales price of $1.46 million.

It is often the case when it comes to money measurements that the median should be used instead of the mean. Another good example is the average salary for individuals working at a large corporation. If the average salary amount that the company reports is the mean rather the median, we should question whether extreme values, like the salaries of executives, were included in the calculation. If they were, we should ask what the median salary is.

We have learned that at the heart of statistical thinking is the concept of the average, a measure of central tendency where most individual values fall in and around this value. However, we have also learned that for certain data, there will be a small number of values far from our measure of central tendency. These values are often called outliers, and it is important that we try to understand why (if possible) they fall so far from the center of the data. For example, a small percentage of patients may respond extremely well to a treatment even though the average treatment response (or effect) was not found to be statistically significant.

An illustrative real-world example is discussed in the *New York Times* article titled "F.D.A. Revokes Approval of Avastin for Use as Breast Cancer Drug." As the article points out, the drug Avastin was approved by the Food and Drug Administration (F.D.A.) as part of an "accelerated program" based on just one study in which patients took Avastin along with another drug named Taxol. However, subsequent studies where Avastin was combined with other chemotherapy drugs did not show Avastin to be effective (on average) at prolonging lives. However, some experts stated that Avastin prolonged the lives of a small number of patients and, therefore, the drug should remain on the market for the sake of those patients.

The distribution of survival times for cancer patients are often skewed to the right or positively skewed. There are often a small percentage of patients who will survive longer that the average survival times of the entire group. In studies such as these, patients are randomized to

NEWS ARTICLE: THE $238 MILLION PENTHOUSE, AND THE HEDGE FUND BILLIONAIRE WHO MAY RARELY LIVE THERE

Web Link: https://www.nytimes.com/2019/01/24/nyregion/238-million-penthouse-sale.html

Search Term: The $238 Million Penthouse

NEWS ARTICLE: F.D.A. REVOKES APPROVAL OF AVASTIN FOR USE AS BREAST CANCER DRUG

Web Link: https://www.nytimes.com/2011/11/19/business/fda-revokes-approval-of-avastin-as-breast-cancer-drug.html?_r=1

Search Term: F.D.A. Revokes Approval of Avastin

treatments to average out possible confounding factors to clearly see whether one treatment is more (or less) effective than another treatment, on average. We are analyzing how effective the treatment is, on average, for all individuals. We are not analyzing how effective the treatment is for a single individual.

When an individual patient's response is far from the average, it is most likely because there are other factors about the patient with values also far from the average. These factors may be contributing to his/her response. Unfortunately, Avastin can't be left on the market for the sake of these patients because there is no way of knowing what these factors are. It would cost the pharmaceutical company a lot of time and money to investigate the characteristics of these patients that led to such long survival times on the treatment. Since it involves such a small percentage of patients (or potential market), the company is unlikely to see it as worth its while.

This is an interesting example of what is known as the tyranny of the average. We have to summarize our data with averages as well as compare averages across treatments in order to decide which treatment is more effective. However, there may be a small percentage of individual values far from the average, known as outliers. If possible, it is important to try to investigate why these individual values are outliers. In the case of Avastin, there could be very interesting reasons why a small percentage of patients reacted so well to the treatment.

4.6 CONCLUSION

We have learned how to summarize the variation contained in our sample of measurements very concisely. When appropriate, we approximated the population distribution of measurements with the normal distribution. When our data follows a normal distribution, we approximated the population mean, a measure of central tendency, with the sample mean. We approximated the population standard deviation, a measure of the spread of values around the mean, with the sample standard deviation.

For distributions of measurements that are considered skewed distributions, we learned that we should use the median and interquartile range, as alternative measures of the central tendency and spread of data values.

We learned how to calculate a z-score that tells us how many standard deviations above or below the mean any particular value lies under the normal distribution.

The normal distribution is also known as a probability distribution. We used this distribution to measure the chance or probability of an individual having a measurement value more extreme than any particular value. In the next chapter, we will ask what does chance mean and learn how we use probability to measure the chance of certain events occurring.

4.7 REAL-WORLD EXERCISES

1. Search the Internet for sample data on a topic of interest to you. Create a histogram of the data describing any interesting features of the data—the interval ranges with the highest and lowest percentage of measurements—the shape of the histogram, and so on. Summarize the data with the appropriate measures of center and spread.

2. Find data that follows the normal distribution or use the data you found for Exercise 1 if it is normal. Calculate the probability that an individual value will be more extreme (or less extreme) than a particular value of your choosing. **Hint:** Calculate the z-score and use the table at https://stattrek.com/online-calculator/normal.aspx to calculate the probability.

3. Think of a type of measurement data (that is not a money measurement) whose distribution you feel is either positively or negatively skewed. Explain your reasons why. Look for an example of the measurement data and create a histogram. Discuss how much the distribution of the data differed from what you expected.

4. Find datasets of interest to you at https://www.kaggle.com/datasets. Choose two quantitative variables from the datasets. Using the Excel AVERAGE and STDEV functions, calculate the mean and standard deviation for both variables. Using the MEDIAN function, calculate the median and interquartile range for both variables. Discuss what each of the sample statistics means in terms of the variable descriptions. Comment on which statistics should be used as measures of central tendency and spread for both variables. **Hint:** To get the interquartile range, order the variable values from smallest to largest. Then, get the median of lower 50% of values (the 1st quartile) and the median of the upper 50% of values (the 3rd quartile).

5. Find a news article that discusses an average. Determine the units of measurements for the variable in question. Is the average discussed in the news article a mean or median? Is the average used in the article the appropriate measure of central tendency for the measurement variable in question? Explain why?

6. Find a journal that provides links to the datasets related to its journal articles. One source is https://www.mdpi.com/journal/data. Select a journal article of interest to you. Download the related datasets. Determine the primary quantitative variable of interest for the analysis, and calculate the sample means across treatment groups. Calculate the sample means at baseline and at end of study if the data is provided for both. Compare those mean values to the values presented in the journal article.

CHAPTER 5

MEASURING UNCERTAINTY WITH PROBABILITY

Probability is expectation founded upon partial knowledge. A perfect acquaintance with all the circumstances affecting the occurrence of an event would change expectation into certainty, and leave neither room nor demand for a theory of probabilities.

—George Boole

5.1 INTRODUCTION

Statistics is the application of mathematics to data for the purpose of turning data into meaningful information about a population of interest. In Chapter 4, we used basic arithmetic to calculate sample statistics like the mean and standard deviation. We calculated the probability of selecting an individual having a measurement value more extreme than a particular value.

Probability is a branch of mathematics that enables us to estimate the chances that an event will occur. It is important that we understand the basic rules of probability because we are often bad at estimating the chances (from our subjective point of view) that an event will occur.

A good example is the birthday experiment. How many people do you think need to be in a room for there to be a 50% chance that two of those people share a birthday? You might be surprised to hear all you need is twenty-three. We will explain why later in the chapter.

KEY TERMS

Probability: A branch of mathematics used for measuring the likelihood of chance events

Confirmation Bias: The tendency to interpret new evidence as confirmation of one's existing beliefs or theories

False Positive: A test result which incorrectly indicates that a particular condition is present

True Positive: A test result which correctly indicates that a particular condition is present

False Negative: A test result which incorrectly detects that a particular condition is not present

True Negative: A test result which correctly indicates that a particular condition is not present

Conditional Probability: The probability of an event given that another event has occurred

Independent Events: Two events, A and B, are independent if the fact that A occurs does not affect the probability of B occurring (and vice versa)

In this chapter, we will ask the following questions:

- How should you think about chance events that happen in your everyday life?
- When you flip a fair coin, why is the probability of a head equal to 0.5?
- How likely do you think it is that a person will win $1 million in the lottery twice on the same day?
- You test positive for a rare disease. Only one in a thousand people in the population have the disease. The test gets it right 99% of the time. What is the probability that you have the disease?
- Why do we only need twenty-three people in a room to have a 50% chance of two of them sharing a birthday?

You will learn some basic rules of probability. You will learn how to look at an event that occurs from a more objective point of view in order to better understand the chances of the event occurring. You will begin to learn about the role chance plays in decision making (regarding the value of population characteristics) through the use of probability.

A VERY LUCKY WIND

Laura Buxton, an English girl just shy of ten years old, didn't realize the strange course her life would take after her red balloon was swept away into the sky. It drifted south over England, bearing a small label that said, "Please send back to Laura Buxton." What happened next is something you just couldn't make up—well, you could, but you'd be accused of being absolutely, completely, appallingly unrealistic.

Web Link: http://www.radiolab.org/story/91686-a-very-lucky-wind/

Search Term: A Very Lucky Wind—Radiolab

5.2 EVERYDAY CHANCE EVENTS

Life is full of incredible events. Is it credible that someone could win $1 million in a lottery twice? How about twice in one day? Incredible, and improbable, yet this happened to a Virginia woman in 2012. Search the Internet and you will find many similar examples. Which, when you consider these events objectively, seem less surprising—tens of millions of people play the lottery every single day. With time, the highly improbable event of someone winning the lottery twice in one day is going to happen. And we often hear of these events because their rarity makes for a good story. In this section we'll consider how we should look at chance events that occur in our everyday lives, by thinking about a true story that happened to a girl from northern England. The story was presented on Radiolab, an online radio show and podcast.

Let's look at the events of this story and see how credible they initially sound (to you). What at first might sound incredible will sound less so when we apply an objective point of view. You'll see that events like these are uncommon, but they do occur; improbable isn't impossible.

So, this girl Laura Buxton released her red balloon that traveled 140 miles southwest, against the prevailing winds, into the backyard of another little girl. Her name was Laura Buxton! The first Laura Buxton wrote a letter to the second Laura Buxton. They eventually arranged to meet.

It turned out that they had a lot more in common than their names. They both were around the same age and both were wearing a pink sweater and jeans. It gets odder. Both had three-year-old Labrador retrievers, gray rabbits, and identical guinea pigs. They were amazed at what was happening and evidently became great friends. The way in which they met and all the coincidences made this all very meaningful to them.

The story is engaging and gets us thinking about the wonderful experiences that can happen by chance. However, we have to be careful about how we reason with the likelihood of chance events that happen in our own lives and in the lives of others.

The girls (and the radio host exploring their story) got very excited about the chance event that led to the girls meeting, both naturally focused on common traits that would make the event seem even more improbable and special. Later in the podcast, the radio host admitted that he selected out what the girls had in common and ignored what they did not have in common.

This tendency toward seeking significance in random events is human nature. Most of us want to believe that life has meaning, that it is not merely a product of chaos, randomness, and chance. When we experience or hear about improbable events that seem to have meaning, we look for confirming evidence that gives it more meaning or significance. In scientific research, this sort of reasoning is called **confirmation bias**. For the researcher with a hypothesis that they really believe in, the danger is in narrowing one's focus to only look for confirming evidence. In the observational world, if you look hard enough, you will eventually find evidence to fit any hypothesis we have about how the world works.

Of course, from the subjective point of view of the two girls, this chance event is very meaningful. There is absolutely nothing wrong with that as long as they don't try and extend that meaning beyond themselves. The search for meaning is (or should be) a very personal and private pursuit. However, from our objective point of view, we should see their meeting as simply a chance event with no meaning attached. The chances of the event happening are small, but unlikely events do happen all the time, and when they do, we will hear about them.

5.3 WHAT IS CHANCE?

Near the end of the novel *War and Peace*, Leo Tolstoy asks the question "What is chance?" He then illustrates through a simple story that when we have complete knowledge and understanding of the circumstances surrounding any event, we have no reason to resort to a concept such as chance at all:

But What Is Chance? What Is Genius?

The words chance and genius do not denote any really existing thing and therefore cannot be defined. Those words only denote a certain stage of understanding of phenomena. I do not know why a certain event occurs; I think that I cannot know it; so I do not try to know it and I talk about chance. I see a force producing effects beyond the scope of ordinary human agencies; I do not understand why this occurs and I talk of genius.

To a herd of rams, the ram the herdsman drives each evening into a special enclosure to feed and that becomes twice as fat as the others must seem to be a genius. And it must appear an astonishing conjunction of genius with a whole series of extraordinary chances that this ram, who instead of getting into the general fold every evening goes into a special enclosure where there are oats—that this very ram, swelling with fat, is killed for meat.

(Citation: Louise and Aylmer Maude, Simon & Schuster, 1942)

From the rams' subjective point of view, they need to resort to chance to understand why the one ram they are observing is "swelling with fat." They do not have full knowledge and understanding of the event they are observing. From our objective point of view, we have full knowledge and understanding of the event, so we do not have to resort to chance to explain it.

When you throw a stone into a pond, a naive observer might credit chance as controlling where it lands. But a more objective observer, if they know the force and angle at which the stone was thrown, could calculate the stone's landing spot precisely. As long as no external forces are at play (like a very windy day), we can predict the event's outcome; chance is not a factor.

The same is true when we flip a coin. In an interesting video titled "How Random Is a Coin Toss?" Professor Persi Diaconis from Stanford University discusses how we could determine the outcome of a coin toss. According to Newton's Laws of Motion, if we could measure how fast the coin was traveling through the air and how fast it was turning over (revolutions per second), we could predict whether it will land heads or tails. In other words, if we could measure the factors or variables (speed of the coin, revolutions per second) that determine the outcome, we could predict the outcome without having to resort to chance. While this is possible in theory, it would be very difficult to do in practice.

In the real world, we don't have complete knowledge and understanding of the many factors that affect an event we observe. For example, if we could measure all of the factors that affect an individual's cholesterol level (and by how much each factor increases or decreases it), we could determine what the individual's cholesterol level will be without having to measure it. However, it is impossible for us to do that. In the observational

HOW RANDOM IS A COIN TOSS?

Web Link: https://www.youtube.com/watch?v=AYnJv68T3MM

Search Term: How Random Is a Coin Toss?

world, how factors (or variables) act (and interact), resulting in an effect of some kind, is often extremely chaotic and complex.

The tools of statistical analysis we will learn in this book enable us to reason with the complex, chaotic and uncertain nature of the world using data and statistics. The statistical models we build are not perfect reflections of reality as it is, but they can be useful for determining relationships between factors and for making predictions. As we have already seen, a well-conducted randomized experiment is a powerful way of averaging out confounding factors in order to be able to make cause and effect conclusions. However, even if we find a statistically significant treatment effect from a randomized experiment, it is important to keep in mind that any estimate regarding the treatment's effectiveness is just that—an estimate. Reality requires working with constraints; we observe only a subset of our target population; and since we do not have complete knowledge and understanding of the population, we have to resort to chance in making a decision about the true effect of the treatment on the entire population. A decision informed by statistics is a decision based in chance.

5.4 MEASURING THE CHANCES OF AN EVENT

We measure the chance of an event occurring with the mathematics of probability. The probability of an event will always be a value between 0 and 1. If the probability of an event is 0, then the event will not occur. If the probability is 1, then the event will occur. Values between 0 and 1 are a measure of the level of uncertainty we have regarding whether or not the event will occur. For example, a probability of 0.5 means there is a 50/50, or 50% chance of the event occurring.

If we were to do a simple experiment of flipping a fair coin one time, what is the probability the coin will land heads up? Most of you would say the answer is 0.5, or 50%. That is the correct answer. But why? Some of you would say it is because the coin has two sides. However, in terms of probability, this is not the reason why!

The probability of a head occurring is not based on the fact that there are two possible outcomes, heads or tails. If that were the case, any event where there are two possible outcomes (rain or not rain tomorrow for example) would be 50/50. It is based on the frequency (or proportion of times) we get a head from repeatedly tossing the coin over and over again. The probability is estimated by taking the number of heads you get out of the total number of coin tosses. If the coin is fair, then we expect this probability to approach 0.5 or 50% as you increase the number of coin tosses.

The probability of any event is the proportion of times the event happens in the very long run, the infinite to be exact. We can't toss a coin forever. Therefore, the probability of any event (getting breast cancer, whether a treatment will work for you) is estimated from empirical evidence drawn from the proportion of such events happening over time.

For example, if a drug for migraine headaches was found in clinical trials to work for 60% of individuals in your age group, we would say that there is a 60% chance that it will work for you. Of course, this also depends on how representative the sample used was of the population. The more people we observe using the drug, in clinical trials or when the drug is on the market, the better we can estimate the true probability of its effectiveness.

One very important assumption when calculating the probability of an event occurring is known as the assumption of independence, or the assumption of **independent events**. In the case of the coin-toss experiment, this means that we assume that getting a head on your first coin toss does not affect the probability of getting a head on your second coin toss, and so on. In a research context this means that the outcome measurements from individual to individual in a study should not affect each other.

Independence is also a very important assumption when it comes to thinking about individual measurements and how they were collected. We will discuss the assumption of independence in Chapter 6, when we begin to reason with the framework used for making statistical decisions about the values of population parameters.

5.5 CONDITIONAL PROBABILITY, BAYE'S RULE, AND THE MAMMOGRAM CONTROVERSY

Every probability that we calculate should be thought of as a **conditional probability**, the probability of an event, given that another event has occurred. This is because the probability of any event is always conditional (or dependent) on some other event(s). In the coin-toss example, the probability of getting a head is conditional on the fairness of the coin. The probability that you get a particular disease is conditional on many factors. For example, the probability that you will get breast cancer is conditional on family genetics and whether you are a man or a woman.

One decision most women make in middle age is when to start mammogram screening for breast cancer. Our intuition tells us that the earlier a woman starts mammogram screening, the better. However, there is a problem with this sort of testing. When a large percentage of a large population get tested for a disease (without necessarily showing any signs and symptoms), it can lead to a lot of **false positives**, even when the actual test for the disease is very accurate. The rarer the disease, the higher the rate of false positives. This is because a large proportion of the people being tested don't have the disease, and if the test is less than perfect, then this will result in a sizable number of false positives.

There has been a lot of controversy in the media regarding the costs and benefits of mammogram screening and whether or not women should begin screening at age forty or later. The CBS News article titled "The High Costs of 'Breast Cancer' False Positives", discusses the costs and benefits of mammogram screening.

The article states that $2.8 billion of the unnecessary medical costs (due to mammogram screening) are the result of false positives, and $1.2 billion can be attributed to over-diagnosis. It goes on to discuss the continuing debate as to when women should start breast cancer screening.

The National Cancer Institute explains the benefits and harms of mammogram screening on their website, including the problem of over-diagnosis and the high rate of false positives.

Breast Cancer Screening

Diagnosis of cancers that would otherwise never have caused symptoms or death in a woman's lifetime can expose a woman to the immediate risks of therapy (surgical deformity or toxicities from radiation therapy, hormone therapy, or chemotherapy), late sequelae (lymphedema), and late effects of therapeutic radiation (new cancers, scarring, or cardiac toxicity).

…

On average, 10% of women will be recalled from each screening examination for further testing, and only 5 of the 100 women recalled will have cancer. Approximately 50% of women screened annually for 10 years in the United States will experience a false positive, of whom 7% to 17% will have biopsies.

It is clear that the financial, emotional, and physical costs of this sort of testing are very high (compared to its benefits) for certain individuals because of the high rate of false positives. So why is there such a high rate of false positives when it comes to this sort of testing?

The answer to this question can be challenging to understand because it is so counterintuitive. It requires stepping back from our subjective view of the result of an individual test, to the objective view of the results for every individual tested. To help us answer this question, we will set up our own testing example. We will imagine randomly selecting 100,000 people to be tested for a rare disease, where the test is accurate but not perfect. We will then calculate the probability than an individual has the disease, given that they tested positive.

The logic we will reason through to calculate this probability is based on what is known as Baye's Rule. It is also known as reversing the conditioning. In other words, we have a known

THE HIGH COST OF "BREAST CANCER" FALSE POSITIVES

Sharpening a medical debate about the costs and benefits of cancer screening, a new report estimates that the US spends $4 billion a year on unnecessary medical costs due to mammograms that generate false alarms, and on treatment of certain breast tumors unlikely to cause problems.

Web Link : http://www.cbsnews.com/news/the-cost-of-breast-cancer-false-positives/

Search Term: The High Cost of Breast Cancer "False Positives"

BREAST CANCER SCREENING

Web Link: http://www.cancer.gov/types/breast/hp/breast-screening-pdq

Search Term: Breast Cancer Screening

effect, a person tests positive for breast cancer. Can we reason in reverse, with the logic of conditional probability, to determine probable cause: the probability the person has the disease given they tested positive.

Let's say we know the probability that any individual has a rare disease is one in a thousand, or 0.001. The test for detecting whether or not an individual has the disease is very accurate, getting it right 99% of the time. When tested, 99% of individuals who have the disease will be found to have the disease, called a **true positive**, whereas 99% of individuals who do not have the disease will be found to not have the disease, called a **true negative**.

You are not showing any signs and symptoms for the disease, but to be on the safe side, you decide to get tested. You test positive for the disease. How concerned should you be? You may be surprised (and somewhat relieved) to learn that the probability you have the disease is only 0.09, or 9%. We will explain how this probability is calculated using Table 5.1.

Table 5.1: Breakdown of Actual Status versus Test Status for a Rare Disease

	Test shows positive	**Test shows negative**	**Total**
Actually sick	99	1	100
Actually healthy	999	98,901	99,900
Total	1,098	98,902	100,000

As already stated, let's say 100,000 individuals are randomly selected from the population and tested for the rare disease. If 1 in every 1,000 individuals has the disease, then we would expect 100 of the individuals in our sample to have the disease. The test for the disease gets it right 99% of the time, so we expect 99 of the 100 individuals with the disease to have a true positive result.

What about the remaining 99,900 individuals who don't have the disease? The test gets it right 99% of the time (and therefore wrong 1% of the time), so we expect 1%, or 999 of these individuals, to have a false positive result.

So in our sample of 100,000 individuals, we will have a total of 1,098 positive results, 99 of which are true positives and 999 are false positives. Therefore, only 99/1098, or 9% of the positive results, are actually true positives. In other words, only 9% of the people who test positive actually have the disease. This answer makes sense when we break it down (as shown in Table 5.1), but it can take some time to fully grasp. Let's try explaining it another way.

You are told that the test gets it right 99% of the time. If you test positive, your reasoning might tell you that you almost certainly have the disease. When we hear that a test result is positive, we have a tendency to focus on the number of people with the disease, and the chances of the test detecting that they have the disease. There are 100 people with the disease, and 99 of those people will be found (by the test) to have the disease. As a result, you may think that your positive result means that there is a 99% chance of you having the disease. The question you are asking yourself is; what are the chances of a positive result, given that an individual (like yourself) has the disease? The answer to this question is there is a 99% chance of this occurring.

However, that is not the question we are trying to answer. We are asking what is the probability that an individual has the disease, given they test positive. To answer this question, we need to think about the number of people who have the disease that test positive, plus the number of people without the disease that test positive.

Remember that the probability of an event is the proportion of times an event occurs over the long run (or repeated observations). In this example, the event of interest is a "true positive" result from testing many individuals over time. We are asking what is the probability an individual has the disease (a true positive), given they test positive. We want to know what proportion of all positive results are true positives.

To answer this question, we need to step back (and take into account) how often the test will come up with a positive result for those individuals who don't have the disease. In this example, it is a very rare disease, so the vast majority of the 100,000 individuals being tested don't have the disease. The test is very accurate, but it still has an error rate of 1%. Therefore, 1%, or 999 of the 99,900 individuals who do not have the disease, will have a false positive result.

Now we know the total number of positive results (1,098) to expect: 99 true positives and 999 false positives. The proportion of these positive results that are true positives is 99/1,098, which is equal to 0.09, or 9%. There is a 9% chance that someone who tests positive actually has the disease.

You can now begin to appreciate the controversy over mammogram screening. A large proportion of the population is being tested for a relatively rare disease without necessarily showing any signs or symptoms. The vast majority of those being tested don't have the disease, so this will result in a large number of false positives as compared to true positives.

The prevalence of false positives led the American Cancer Society to recommend that women start mammogram screening later and less frequently. This recommendation is discussed in the *New York Times* article titled "American Cancer Society, In a Shift, Recommends Fewer Mammograms."

The American Cancer Society did make it clear that women are free to make their own decisions when to begin screening. Remember the calculations in our example were based on the assumption the individuals were not showing any signs and symptoms. Every woman will have her own risk factors and reasons (genetics and family history for example) for wanting to have a mammogram. However, her doctor should communicate clearly the chances of getting a false positive and what further (and perhaps invasive) procedures a positive result could lead to.

5.6 A TRAGIC STATISTICAL ERROR—THE CASE OF SALLY CLARK

We will now look at the tragic case of Sally Clark. In September 1996, her first child died a few weeks after he was born. In December 1998, her second child died. She was accused of murdering her children, and justice was laid down quite swiftly with her conviction for murder in November 1999. The tragedy was that she did not murder her children, and the conviction was based on very misleading statistics. The article from the *Telegraph*, titled "Misleading Statistics Were Presented as Facts in Sally Clark Trial" describes how a lack of understanding of the statistics presented during the case led to the wrongful conviction of Sally Clark.

When two events, A and B, are considered independent, then the probability of both events occurring is calculated by multiplying the probabilities of each event together. This can be written as follows:

$$\text{Prob}(A \text{ and } B) = \text{Prob}(A) \times \text{Prob}(B)$$

AMERICAN CANCER SOCIETY, IN A SHIFT, RECOMMENDS FEWER MAMMOGRAMS

One of the most respected and influential groups in the continuing breast-cancer screening debate said on Tuesday that women should begin mammograms later and have them less frequently than it had long advocated.

The American Cancer Society, which has for years taken the most aggressive approach to screening, issued new guidelines on Tuesday, recommending that women with an average risk of breast cancer start having mammograms at 45 and continue once a year until 54, then every other year for as long as they are healthy and likely to live another 10 years.

Web Link: http://www.nytimes.com/2015/10/21/health/breast-cancer-screening-guidelines.html

Search Term: American Cancer Society, in a Shift, Recommends Fewer Mammograms

MISLEADING STATISTICS WERE PRESENTED AS FACTS IN SALLY CLARK TRIAL

The shadow of Sally Clark's wrongful conviction for murdering two of her babies, and the use of misleading statistics by expert witnesses, will have hung heavy over the jury.

The evidence in the 1999 trial was so badly flawed that the Royal Statistical Society of London wrote to the Lord Chancellor in concern and suggested that courts use expert statistical witnesses to avoid a similar debacle. The case also raised concerns about the way that statistics are handled in court. The Sally Clark jury was told by Prof Sir Roy Meadow that the chances of a mother losing two babies to sudden infant death syndrome were "one in 73 million". The figure was compiled from national child mortality statistics.

Web Link: http://www.telegraph.co.uk/news/uknews/1432762/Misleading-statistics-were-presented-as-facts-in-Sally-Clark-trial.html

Search Term: Misleading Statistics Were Presented as Facts in Sally Clark Trial

The expert witness in this case was Professor Sir Roy Meadow, a retired pediatrician. He found a study that stated the chances of a professional, nonsmoking family experiencing a cot death or Sudden Infant Death Syndrome (SIDS) was 1/8,500, or 0.0001. Meadow calculated that the chances of a family losing two babies to SIDS was "one in 73 million," or 0.00000001. Meadow arrived at this probability by assuming the two events were independent. He multiplied the probability of a single cot death by itself.

Let A be the event that Sally Clark's first child dies from SIDS. Let B be the event that Sally Clark's second child dies from SIDS.

$$\text{Prob}(A \text{ and } B) = 0.0001 \times 0.0001$$
$$= 0.00000001$$

As the Mayo Clinic points out on its website in an article titled "Sudden Infant Death Syndrome (SIDS)," the likely factors associated with sudden infant death are both physical (brain abnormalities, low birth weight, respiratory infection) and sleep environmental (sleeping on the stomach or side, sleeping on a soft surface, sleeping with parents). Both children were born to the same parents in the same environment, greatly increasing the probability that their second child could die from SIDS, given that their first child died in this way. The assumption that these two tragic events were independent was not valid in this case.

The probability of a family experiencing two sudden infant deaths is still very low. However, we need to remember that probability is the proportion of times an event happens over the long run. With so many children being born every year, it is going to happen to some family, somewhere, sometime. Unfortunately for the Clark family it happened to them.

SUDDEN INFANT DEATH SYNDROME (SIDS)

Web Link: https://www.mayoclinic.org/diseases-conditions/sudden-infant-death-syndrome/symptoms-causes/syc-20352800

Search Term: Diseases and Conditions Sudden Infant Death Syndrome (SIDS)

5.7 THE BIRTHDAY EXPERIMENT

We began this chapter with mentioning the birthday experiment that states we only need twenty-three people in a room to have a 50% chance that two people will share the same birthday! If that number seems too low, it's because you're thinking about the problem from your subjective point of view. How many people do we need in a room so that someone shares a birthday with me? The number starts to make more sense when you try to view the problem from an objective point of view.

With twenty-three people in a room, the first person can make a one-to-one connection with twenty-two other people. The second person has connected with the first person, but there

> **HOW COMMON IS YOUR BIRTHDAY?**
>
>
>
> Web Link: http://www.nytimes.com/2006/12/19/business/20leonhardt-table.html?_r=1
>
> Search Term: How Common Is Your Birthday?
>
> **HOW COMMON IS YOUR BIRTHDAY? —VIZWIZ**
>
>
>
> Web Link: http://www.vizwiz.com/2012/05/how-common-is-your-birthday-find-out.html
>
> Search Term: How Common Is Your Birthday?

are twenty-one other people left in the room that he or she can connect with. The third person has connected with the first and second persons, but there are twenty other people left in the room that he or she can connect with.

We can continue this logic going from person to person, adding up the number of one-to-one connections made, until we get to the twenty-third person. We find that there is a total of 253 one-to-one connections we can make between twenty-three people. Is it so surprising that there is a 50% chance that one of these pairings will share a birthday?

The 50% chance is calculated on the assumption the same number of children are born on any given day of the year. In reality, this is not the case. If you were to try this experiment, there is a good chance that you will find two people with matching birthdays in September, a very busy month for births. According to a *New York Times* article titled "How Common is Your Birthday?" (looking at birth dates of children born from 1973 to 1999), the most common birthday is September 16th. If you were born around that time of year, do the math, and ask your parents the circumstances around which you were conceived. It could lead to an interesting conversation. You can find an interactive graphic showing the distribution of the most common to least common birth dates on a website called Vizwiz.

5.8 CONCLUSION

In this chapter, we have learned how to reason with the chance of an event occurring from an objective point of view. We learned that the probability of any event occurring is calculated by observing the proportion of times the event happens over time. For the medical testing example, we found the probability a person who tests positive has the disease by calculating the proportion of all positive results that were true positives.

It is this mindset of stepping back to view the big picture of possibilities that is required for making good statistical decisions. In the next chapter, we will start to reason with sampling variation: how sample statistics vary from sample to sample. We will step back and think about all the possible sample statistics we could get (based on a given sample size) when we assume the population parameter of interest is equal to a particular value.

5.9 REAL-WORLD EXERCISES

1. Ask three of your friends to tell you about a surprising chance event that happened in their lives. How surprising did you find the events your friends described?

2. Flip a fair coin a hundred times, recording the proportion of heads you got after every ten coin flips. After every ten coin flips, comment on the difference between the proportion of heads you got and the proportion that you expected.

3. Do an Internet search to find out the chances of someone in a specific demographic (age group, gender, ethnicity) getting a particular disease. Calculate the probability that a person who tests positive for the disease actually has the disease. Base your calculations on three different accuracy levels for the medical test: 90%, 95%, and 99%.

4. Think of events that happen in your everyday life—how much you eat, sleep, work, walk, exercise, suffer illnesses, commute, talk with friends, use social media, etc. Choose two events you feel are independent and two events you feel are not independent. Explain why?

5. A meteorologist might state that there is a 38% chance of rain tomorrow. What do think this statement really means? What is the probability that it won't rain tomorrow?

6. Do an Internet search to find out the estimated percentage of people who are left-handed in the US. If three people are selected at random from the US population, what is the probability that:

    a. All three will be left-handed

    b. All three will be right-handed

    c. One of the three will be left-handed

    d. Two of the three will be left-handed

    Hint: Assume independence between individuals as to whether or not they are left- or right-handed. When thinking about c) keep in mind that any one of the three people selected could be left-handed. You have to calculate the probability for each possible scenario and add them together.

VISUALIZING AND SUMMARIZING SAMPLE STATISTICS

CHAPTER 6

I know of scarcely anything so apt to impress the imagination as the wonderful form of cosmic order expressed by the "Law of Frequency of Error". The law would have been personified by the Greeks and deified, if they had known of it. It reigns with serenity and in complete self-effacement, amidst the wildest confusion. The huger the mob, and the greater the apparent anarchy, the more perfect is its sway. It is the supreme law of Unreason. Whenever a large sample of chaotic elements are taken in hand and marshalled in the order of their magnitude, an unsuspected and most beautiful form of regularity proves to have been latent all along.

—Sir Francis Galton, *referring to the central limit theorem*

6.1 INTRODUCTION

In Chapter 4, we learned how to reason with and describe the variation in individual measurements. We saw that many measurements in nature (such as height and blood pressure) tend to be distributed according to the normal distribution. We described the variation in individual measurements around the mean using the standard deviation and the empirical rule for the normal distribution.

KEY TERMS

Statistical Decision: A decision based in chance regarding the value of a population parameter

Sample Mean: A measure of central tendency for a single sample of quantitative data

Population Mean: A measure of central tendency for a single population of quantitative data

Sample Mean Difference: The difference between two sample means for quantitative data

Population Mean Difference: The difference between two population means for quantitative data

Sample Proportion: A statistic representing a proportion for a single sample of categorical data. Example: the number of voters in a sample who will vote for a candidate over the sample size of voters

Population Proportion: A parameter representing a proportion for a single population of categorical data. Example: the number of voters in a population who will vote for a candidate over the total population size of voters

Sample Proportion Difference: The difference between two sample proportions for categorical data

Population Proportion Difference: The difference between two population proportions for categorical data

Sampling Variation: The variation in statistics from sample to sample

Sampling Distribution: The name given to the distribution of sample statistics

Independent Measurements: The probability that any individual (in the sample) has a particular measurement value should not be affected by the measurement value of other individuals in the sample. Less technically, it means that individual measurements have no effect on each other

Standard Error: A measure of how far we expect a sample statistic to deviate (on average) from the population parameter of interest

Central Limit Theorem: The distribution of sample statistics (from numerous random samples of the same size) will be normally distributed and centered at the population parameter of interest

Hypothesis Testing: A method of statistical analysis used for making statistical decisions regarding the values of population parameters

Null Hypothesis: A general statement or default position regarding the value of a population parameter. When comparing treatments, it is a statement that the treatments are equally effective

Null Value: The value for the population parameter presented in the null hypothesis

Confidence Interval: An interval of plausible values for the population parameter of interest

In this chapter, we will begin to reason with and describe variation in sample statistics by asking the following questions:

- What do we mean by variation in sample statistics?
- How far do we expect our sample statistic to be from the population parameter of interest?
- How are statistics from different samples distributed?
- How do we measure variation in sample statistics?

We will learn how to reason with and measure variation in sample statistics. We will learn how the variation in sample statistics follow a familiar pattern – the normal distribution. We will learn that the normal distribution of possible sample statistics is centered at the unknown population parameter of interest. In other words, the average of all possible sample statistics is the truth. This result, known as the **Central Limit Theorem**, gives us a framework for making **statistical decisions** about the population parameter of interest. We will learn how to measure how far we expect our sample statistic to deviate from the population parameter with what is known as the **standard error**.

The concepts we will study over the next three chapters are the foundations of statistical inference (or analysis). Laying down foundations are never easy in any context, so it is important we reason with the concepts slowly and thoughtfully, one step at a time. We will reason through the same process for four primary population parameters of interest. The process will feel repetitive, but this repetition helps to engrain the concepts and ensure that your understanding of statistical analysis is based on firm foundations.

6.2 THE DISTRIBUTION OF SAMPLE STATISTICS

Let's say you want to estimate the **population mean** height of men at your college. You decide to collect a random sample of 16 men's heights. The **sample mean** height you obtain turns out to be 69.2 inches.

You are interested in understanding variation in sample statistics, so you decide to collect another sample of 16 men's heights. This time the sample mean height is equal to 70.5 inches. You are not surprised that the sample means are different in each sample because they are based on a different sample of individuals. However, you are interested in understanding how much they will vary if you were to repeat this process of sampling numerous times. You decide to repeat the process ten more times, resulting in the following 12 sample mean heights:

68.3, 68.7, 69.2, 69.4, 69.6, 69.9, 70.1, 70.3, 70.5, 70.9, 71.1, 71.4

At this stage, some questions might come to mind:

- Which of the 12 sample mean heights is the best estimate of the population mean height of men at your college?

If you selected a random sample each time, all 12 sample mean heights are equally valid estimates of the population mean height.

- Why does the range of sample mean heights go from 68.3 inches to 71.4 inches?

In reality, the 12 sample mean heights you observe (based on just 16 individual men's heights in each case) would not have a range going from 68.3 inches to 71.4 inches. However, if you were to repeat this process a very large number of times, this is a probable range within which you would expect most of your sample means to fall. You will learn why and the assumptions we had to make for this to be true.

Figure 6.1 is a histogram of sample mean heights (resulting from 10,000 samples) each using a sample size of 16 men. It was generated (or simulated) using statistical software. The software generates a random sample of 16 individual heights, calculates the sample mean height, and repeats this process 10,000 times. It then takes the 10,000 resulting sample mean heights and constructs a histogram of sample means.

We can see from Figure 6.1 that the vast majority of sample means will be between 68.3 and 71.4 inches. To be more precise, we expect 95% of sample means will be between 68.5 inches and 71.5 inches. However, in order for this to be true, we had to make certain assumptions.

Every population will have its own mean and standard deviation. Our sample mean and sample standard deviation are estimates of those population parameters. In order to be able to create and visualize the **sampling distribution** of possible sample means shown in Figure 6.1, we had to tell the computer software that the population mean height is equal to 70 inches and the population standard deviation is equal to 3 inches.

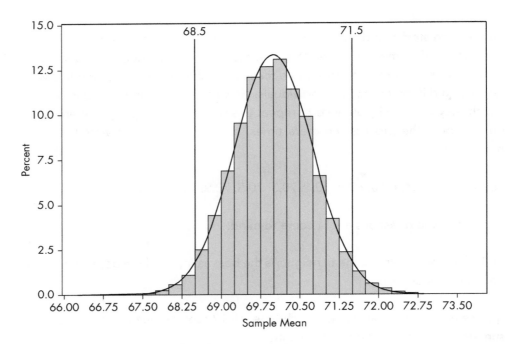

Figure 6.1: Simulation of 10,000 Sample Means (Sample Sizes equal to 16 Men)

In other words, in order to be able to visualize this distribution, we have to assume that the population mean and population standard deviation are equal to particular values. In Chapter 8, when making statistical decisions regarding the true value of the population mean using **hypothesis testing**, we will start with an assumed value for the population mean. This will enable us to construct and visualize the sampling distribution of possible sample means (based on a particular sample size), and therefore determine the likelihood of obtaining our sample mean, given the assumed value for the population mean is correct.

The distribution of possible sample mean heights is known as the sampling distribution. In the next section, we will use what is known as the standard error as a measure of variation in possible sample mean heights. It includes the standard deviation and sample size in its calculation. We will see that the larger the sample size, the smaller the standard error, and therefore the narrower the spread or variation in possible sample means based on a particular sample size.

No matter what the value of the population mean is, the distribution of sample mean heights (based on a particular sample size) will follow the normal distribution, centered at the population mean height. However, this result is only valid when the data collected adheres to the following necessary assumptions and conditions:

1. **Random Sample**: A sample that is representative of the population

2. **Independence**: The individual measurements have no effect on each other

3. **Independent Groups**: Measurements across groups have no effect on each other
4. **Sample Sizes**: The sample sizes are sufficiently large

A random sample is representative of the population, but should also give us **independent measurements**. For example, in a diet study, an individual's weight measurement should not be dependent on another individual's weight measurement. If you selected a sample with a husband and wife, sharing similar eating habits, their weight measurements would be dependent on each other. A random sample should ensure that this sort of situation does not occur. However, it is also important that during the study, individuals are not interacting with each other in a way that would affect each other's weight measurements. When comparing groups (or treatments), it is an important assumption that the measurements are also independent of each other across groups. For a diet study comparing weight loss of males and females, a sample with a husband and wife who shared similar eating habits would result in weight measurements that are dependent across gender.

Larger random sample sizes are always better at estimating population parameters. However, if the sample size is too large relative to the population size, the assumption of independence between measurement values may also break down. Please see the appendix for an explanation with example of how this situation can occur. The example describes how independence (as a probability rule) between measurements can break down when the sample size is too large. As a rule of thumb, as long as the sample size is no more than 10% of the population size, independence between measurements should be a valid assumption. Later in this chapter, we will discuss the sample size condition and how large a sample size is large enough for the result of central limit theorem to be valid.

The population parameter is still unknown but the result of the central limit theorem enables us to describe **sampling variation**, the variation in sample statistics from sample to sample. Reasoning with sampling variation will enable us to make a statistical decision regarding the value of the population parameter.

Technically, for quantitative data, the sample statistics (like sample means) follow what is known as the t-distribution and not the normal distribution. There is a unique t-distribution for every sample size. When making statistical decisions regarding the value of a population mean using a small sample size, our calculations will be more precise when we use the t-distribution rather than the normal distribution. The t-distribution has got fatter tails (the ends of the curve) than the normal distribution when working with small sample sizes, thus affecting the calculations used for making statistical decisions.

However, for sample sizes of 60 or more, these two distributions are aproximately the same distribution. For ease of explanation, when discussing the distribution of sample statistics, based on quantitative data, we will simply state that the sample statistics follow the normal distribution. Please see appendix for an in-depth explanation of the t-distribution.

6.3 MEASURING VARIATION IN SAMPLE STATISTICS

In this chapter, we will reason with a framework for making statistical decisions that is the result of the central limit theorem for the following four primary population parameters of interest:

1. Population mean for quantitative data. Examples: weight, height, blood pressure, etc.
2. Population mean difference between two groups for quantitative data
3. Population proportion (or percentage) for categorical data. Example: The proportion or percentage of voters who will vote for a candidate in an election
4. Population proportion difference between groups for categorical data

We will measure the variation in sample statistics using what is known as the standard error. It measures how far on average we expect a sample statistic to deviate from the unknown population parameter. We will discuss what the standard error means and look at how it is calculated for each of the four population parameters of interest.

The meaning of the standard error is the same for the four population parameters of interest, though its calculation is different in each case. To allow us to focus on understanding its meaning, you will find most of the standard error calculations for each of our examples in the appendix.

6.3.1 Population Mean

The population mean is the measure of central tendency for a single population of quantitative data. In Chapter 4, we measured the variation in individual measurement values around the sample mean using what what is known as the sample standard deviation. It measures how far (on average) we expect any individual value to deviate from the sample mean. We measure the variation in sample means around the population mean using the standard error. It measures how far (on average) we expect a sample mean to deviate from the population mean.

The standard error is calculated by dividing the standard deviation by the square root of the sample size as follows:

$$se(\bar{x}) = \frac{s}{\sqrt{n}}$$

Standard error = standard deviation/square root of sample size

Ideally, we would like to use the population standard deviation in this calculation. However, since this value (like the population mean) is unknown, the sample standard deviation is used as an estimate of this value.

The sample standard deviation is above the division line, or what is known as the numerator in the calculation of the standard error. The square root of the sample size is below the division line, or what is known as the denominator. As a result, the larger the sample size, the smaller the standard error, and therefore the narrower the range of possible sample means we expect to obtain.

This should make intuitive sense. If we select a random sample of men, the larger the sample size, the more data (or information) our sample mean height is based on. A random sample should ensure we obtain a representative sample of men's heights. The larger the sample size, the closer we expect our sample mean height to be to the population mean height.

Example 6.1

Let's return to our (hypothetical) example of men's height in the United States from Section 6.2. In order to be able to visualize the normal distribution of sample mean heights (based on a particular sample size) around the population mean height of men in the United States, we will use the currently accepted value for the population mean.

According to the top Google search result (from halls.md) in 2019, the currently accepted population mean height of men in the United States is equal to 70 inches. The value for the standard deviation is not given, but 3 inches is considered a good estimate. In the NHANES men's height data, the sample standard deviation was equal to 7.7 cm, which is equal to 3 inches. The population distribution would look like Figure 6.2.

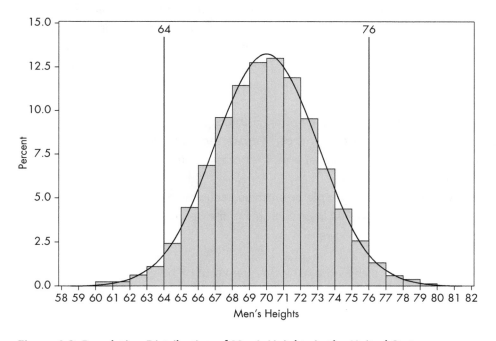

Figure 6.2: Population Distribution of Men's Heights in the United States

The standard deviation of 3 inches measures the average distance we expect an individual height value to be from the population mean. From Figure 6.2 and the empirical rule for the normal distribution (discussed in Chapter 4), we can state that (approximately) 95% of individual heights lie within two standard deviations of the population mean height of 70 inches, or between 64 inches and 76 inches.

Let's say we select a random sample of 36 men. In what range of values do we expect our sample mean height to fall? To answer this question, we calculate the standard error as follows:

standard error = standard deviation/square root of sample size
= 3/square root of 36
= 3/6
= 0.5 inches

The standard error tells us that we expect a sample mean height (based on 36 individual heights) to deviate by 0.5 inches (on average) from the population mean height. If the population mean height is equal to 70 inches, according to the empirical rule for the normal distribution, 95% of possible sample mean heights should be within two standard errors of the population mean, or between 69 inches and 71 inches. Only 5% of sample mean heights are expected to lie beyond these values. If we obtain a sample mean outside this range of values, it will throw into doubt the assumption that the population mean is equal to 70 inches. This is the type of reasoning we will use when conducting hypothesis testing in Chapter 8.

Using statistical software, we can simulate the selection of numerous samples of 36 men's heights and calculate the sample means. Figure 6.3 shows a histogram of 10,000 sample mean heights (based on samples of 36 men), assuming the population mean height is equal to 70 inches.

Figure 6.3 shows the curve for the normal distribution of possible sample means, constructed using the standard error of 0.5 inches. We can see that the histogram (resulting from the simulation of 10,000 sample means) and the normal distribution (resulting from the central limit theorem) match up very well.

As mentioned previously, sample size drives accuracy in statistical testing. If we were to increase the sample size to 144 men (assuming the same sample standard deviation), the standard error would be as follows:

standard error = 3/square root of 144
= 3/12
= 0.25 inches

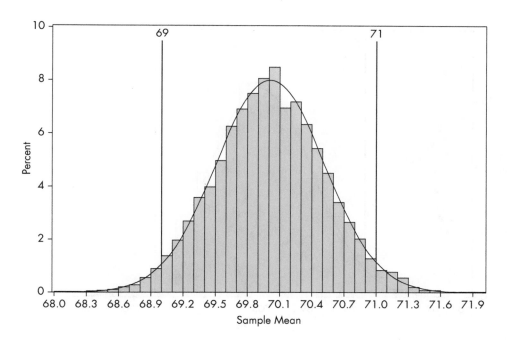

Figure 6.3: Simulation of 10,000 Sample Mean Heights—Sample Sizes Equal to 36 Men

The standard error tells us that we expect a sample mean height (based on 144 individual heights) to deviate by 0.25 inches (on average) from the population mean height. If the population mean is equal to 70 inches, 95% of possible sample means should be within two standard errors of the population mean, or between 69.5 inches and 70.5 inches. Only 5% of sample means are expected to lie beyond these values. If we obtain a sample mean outside this range of values, it will throw into doubt the assumption that the population mean is equal to 70 inches.

The larger the sample size, the smaller the standard error, the narrower the range of possible sample mean heights around the assumed value for the population mean height. However, it should be noted that, due to the square root in the calculation, we had to increase the sample size by four times (from 36 men to 144 men) to cut the standard error in half, from 0.5 to 0.25 inches. In other words, a larger random sample is more likely to result in more precise estimates of population parameter but there are diminishing returns (less bang for your buck) from increasing the sample size.

In the Fast Diet study that we critiqued in Chapter 2, the subjects were randomized to two diets: The Intermittent Energy Restriction (IER) or fast diet, and the Continuous Energy Restriction (CER) or regular diet.

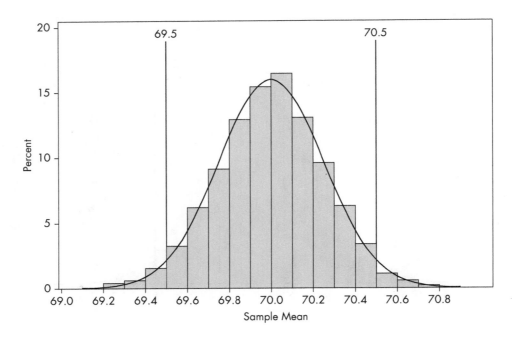

Figure 6.4: Simulation of 10,000 Sample Mean Heights—Sample Sizes Equal to 144 Men

According to the study, after six months, the 53 subjects on the IER diet lost an average of 6.4 kg while the 54 subjects on the CER diet lost an average of 5.6 kg. These sample mean weight losses are estimates of the population mean weight losses on each diet. The researchers used these sample mean weight losses (for both treatments) along with their standard errors (not presented in the journal article) to construct what are known as confidence intervals for the population mean weight losses. We will look at the confidence intervals in our next chapter and explain what they mean.

6.3.2 Population Mean Difference

Let's say we have a sample of patients randomized to two treatments for lowering cholesterol levels. In this case, we are making a comparison of sample mean cholesterol levels. We are interested in whether or not the population mean cholesterol level is different for treatment A compared to treatment B.

The **population mean difference** is the parameter of interest when comparing groups for quantitative data. The difference in the sample means, called the **sample mean difference**, is an estimate of the population mean difference. The terms sample effect size and population effect size are also used. The standard error measures how far (on average) we expect a sample mean difference to deviate from the population mean difference.

Example 6.2

Let's say we are interested in estimating the population mean difference in heights between men and women in the United States. We select a random sample of 36 men and 57 women, a total sample size of 93 individuals. Note that the number of men and women in our sample do not have to be the same.

For the men, we calculate a sample mean height equal to 69 inches and a sample standard deviation equal to 3 inches. For the women, we calculate a sample mean height equal to 65 inches and a sample standard deviation equal to 2.5 inches. The standard error is equal to 0.60 inches (see appendix for calculation).

The standard error tells us that we expect a sample mean difference (men minus women) in height (based on a total sample size of 93 men and women) to deviate by 0.60 inches (on average) from the population mean difference in height.

The sample mean difference of 4 inches (69 inches minus 65 inches) is a sample estimate of the population mean difference. If for example, we were to assume the population mean difference in height is actually 3 inches, we would expect (based on a total sample size of 93 men and women) 95% of sample mean differences in height to be within two standard errors of 3 inches, or between 1.8 inches and 4.2 inches.

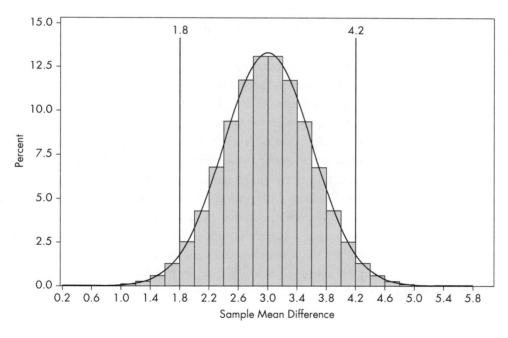

Figure 6.5: Simulation of 10,000 Sample Mean Differences in Height (Based on Total Sample Size of 93 Men and Women)

Therefore, our sample mean difference of 4 inches is a fairly reasonable value to obtain under such an assumption. In other words, it is quite probable (based on a sample of 93 men and women) to obtain a sample mean difference of 4 inches, given that the population mean difference is equal to 3 inches. Again, this is the sort of reasoning we will use when making statistical decisions regarding the value of population parameters using hypothesis testing in Chapter 8.

In the college football study discussed in Chapter 2, the researchers were interested in comparing the mean hippocampal volume across three groups: players with concussion, players without concussion, and a control group. There were 25 subjects in each group. The sample means differences are presented in Table 6.1:

Table 6.1

Comparison Groups	Sample Mean Difference
Controls versus (minus) Players with Concussion	1,788 uL
Controls versus (minus) Players without Concussion	1,027 uL
Players without Concussion versus (minus) Players with Concussion	761 uL

The sample mean differences in hippocampal volume are estimates of the population mean differences. The researchers constructed confidence intervals for the population mean differences using the sample mean differences and their standard errors. We will examine the confidence intervals in our next chapter and explain what they mean.

In the Fast Diet study, the sample mean difference in weight loss (IER minus CER) was 0.8 kg (6.4 minus 5.6). The sample mean difference in weight loss is an estimate of the population mean difference in weight loss. Using the reasoning of hypothesis testing, the researcher asked the question: What is the likelihood of getting a sample mean difference of 0.8 kg (or more extreme) in a sample of 107 women given that the population mean difference is equal to 0 kg? In other words, is the sample mean difference in weight loss of 0.8 kg large enough (or far enough away from 0 kg) for the researcher to make the claim that the IER diet is more effective at reducing weight on average? We will discuss this type of reasoning in greater depth in Chapter 8 along with the results of this hypothesis test.

6.3.3 Population Proportion

The **population proportion** is the parameter of interest when looking at a single sample of categorical data. For example, we might be interested in estimating the proportion of the population who will vote for a particular candidate in a national election. The population proportion is equal to the number of voters in the population who will vote for the candidate over the population size. The **sample proportion** is equal to the number of voters in your sample who will vote for the candidate over the sample size. The sample proportion is an estimate of the population proportion. The standard error measures how far (on average) we expect a sample proportion to deviate from the population proportion.

Example 6.3

The coin-toss experiment can be thought of as a question regarding a population proportion. We learned in Chapter 5 that if we toss a fair coin a very large number of times, we expect the proportion of heads to be 0.50. We can think of this value as the population proportion. To test whether the coin is fair, we can estimate the population proportion by tossing the coin a certain number of times and calculating the sample proportion of heads.

We decide to toss our coin 100 times and calculate the sample proportion of heads. This value is equal to the number of heads we get over the number of coin tosses. Assuming the coin is fair, the standard error of the sample proportion is equal to 0.05 (see appendix for calculation).

The standard error tells us that we expect a sample proportion of heads based on 100 coin tosses to deviate by 0.05 (on average) from the population proportion. If the coin is fair, 95% of possible sample proportions should be within two standard errors of 0.50, or between 0.40 and 0.60. Only 5% of sample proportions are expected to lie beyond these values. If we obtain a sample proportion outside this range of values, it will throw into doubt the assumption that the coin is fair.

We can test the theory by simulating numerous experiments of 100 coin tosses. Figure 6.6 shows a histogram of 10,000 sample proportions, each based on tossing a coin 100 times.

Figure 6.6 also shows the curve for the normal distribution of possible sample proportions constructed using the standard error of 0.05. We can see that the histogram (resulting from 10,000 simulations of sample proportions) and the normal distribution (resulting from the central limit theorem) match up very well.

If we increase the number of coin tosses to 400, the standard error would be equal to 0.025 (see appendix for calculation), half the size it was when we only toss the coin 100 times. As a result (if the coin is fair), 95% of possible sample proportions should lie within a narrower range, between 0.45 and 0.55, compared to the range from 0.40 to 0.60, based on 100 coin tosses. In this case, if we obtain a sample proportion outside this range of values, it will throw into doubt the assumption that the coin is fair.

We can test the theory by simulating the selection of numerous samples of 400 coin tosses and calculating the sample proportion of heads. Figure 6.7 shows the histograms of 10,000 sample proportions based on 400 coin tosses, assuming the coin is fair.

102 | STATISTICAL THINKING THROUGH MEDIA EXAMPLES

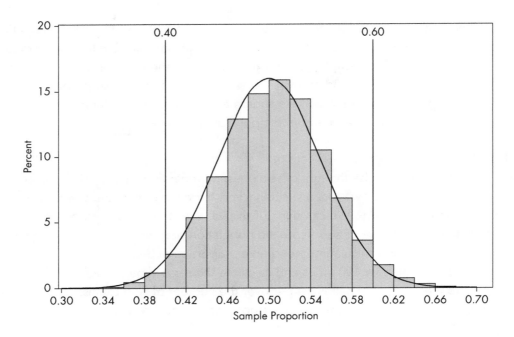

Figure 6.6: Simulation of 10,000 Sample Proportions of Heads (Based on 100 Coin Tosses)

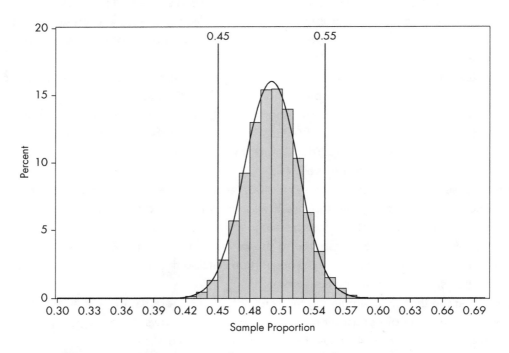

Figure 6.7: Simulation of 10,000 Sample Proportions of Heads (Based on 400 Coin Tosses)

We can see that the histogram (resulting from 10,000 simulations of sample proportions) and the normal distribution (resulting from the central limit theorem) match up very well.

In Chapter 3, we discussed the results of a survey conducted by The Pew Research Center titled "Political Polarization and Media Habits." One of the main findings was that 47% (or 0.47) of Americans whom the report defines as "consistently conservative" named Fox News as their main source for news on politics.

The sample proportion of 0.47 is an estimate of the population proportion. In Chapter 7, we will use the sample proportion along with the standard error to construct a confidence interval for the population proportion of consistent conservatives who named Fox News as their main source for news on politics.

6.3.4 Population Proportion Difference

When comparing groups for categorical data, say, a sample of male and female voters in an election on whether or not they plan to vote for a particular candidate, we make a comparison of sample proportions. We are interested in whether or not the proportion of males in the population, who are planning to vote for the candidate, is different from the proportion of females who are planning to do so.

The **population proportion difference** is the parameter of interest when comparing groups for categorical data. It is the difference between the population proportions for both groups. The **sample proportion difference** is an estimate of the population proportion difference. The standard error measures how far we expect (on average) a sample proportion difference to deviate from the population proportion difference.

Example 6.4

Let's say that we are interested in whether or not the proportion of males in the population (who plan to vote for a particular candidate) is different from the proportion of females. We select a random sample of 100 men and 100 women. Note that the sample sizes are the same, but they do not have to be. We found that 55 of the 100 men said they would vote for the candidate, a sample proportion equal to 0.55. We found that 45 of the 100 women said they would vote for the candidate, a sample proportion equal to 0.45.

The sample proportion difference of 0.10 (0.55 minus 0.45) is a sample estimate of the population proportion difference between men and women. The standard error of the sample proportion difference is equal to 0.07 (see appendix for calculation). This value tells us that we expect a sample proportion difference (based on a total sample size of 200 men and women) to deviate by 0.07 (on average) from the population proportion difference.

Let's assume that the population proportion difference is in fact 0.20 (meaning 20% more men than women will vote for the candidate). If this assumption is true, 95% of sample proportion differences should be within two standards errors of 0.20, or between 0.06 and 0.34 (see Figure 6.8). Only 5% of sample proportion differences are expected to lie beyond these values. Therefore, our sample proportion difference of 0.10 (based on a total sample size of 200 men and women) is a reasonable value to expect under this assumption.

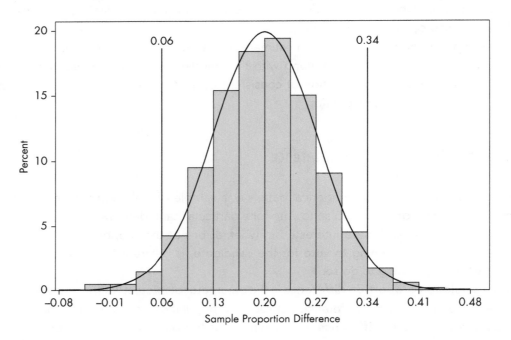

Figure 6.8: Simulation of 10,000 Sample Proportion Differences Based on Total Sample Size of 200 Men and Women

Let's say we increase the sample size to 400 men and 400 women and obtained the same sample proportion difference of 0.10. The standard error in this case is equal to 0.035 (see appendix for calculation). A sample proportion difference, based on sample sizes of 400 in each group, is expected to deviate (on average) from the population proportion by 0.035.

Again, let's assume that the population proportion difference is 0.20. If this assumption is true, 95% of sample proportion differences (based on a total sample size of 800 men and women) should be within two standard errors of 0.20, or between 0.13 and 0.27 (see Figure 6.9). Only 5% of sample proportion differences are expected to lie beyond these values. In this case, our sample proportion difference of 0.10 is not a reasonable value to expect under the assumption the population proportion difference is equal to 0.20.

Using the larger sample size, we expect a narrower range of possible sample proportion differences centered at the assumed value for the population proportion difference of 0.20. In this case, obtaining a sample proportion difference of 0.10 (or 10%) throws into doubt the assumption that the population proportion difference is equal to 0.20 (or 20%). Our sample estimate is more than two standard errors away from what we assumed the population parameter to be.

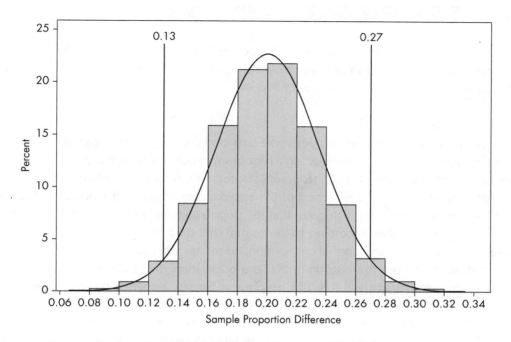

Figure 6.9: Simulation of 10,000 Sample Proportion Differences Based on Total Sample Size of 800 Men and Women

In the Political Polarization & Media Habits survey, two other interesting findings were as follows:

1. Consistent conservatives (CC) see more Facebook posts in line with their views than consistent liberals (CL), 47% versus 32%, respectively.

2. Consistent liberals (CL) are more likely to block others because of politics than consistent conservatives (CC), 44% versus 32%, respectively.

For 1) above, the sample proportion difference (CC minus CL) of 0.15 (0.47 minus 0.32) is an estimate of the population proportion difference. For 2) above, the sample proportion difference (CL minus CC) of 0.12 (0.44 minus 0.32) is an estimate of the population proportion difference.

In Chapter 7, we will use the sample proportion differences along with the standard error to construct confidence intervals for the population proportion differences for both of these findings. In Chapter 8, we will conduct hypothesis tests to determine whether the sample proportion differences provide enough evidence to conclude that the population proportion differences are not equal to zero. In other words, do consistent conservatives and consistent liberals in the US population truly differ (in their opinions) in the way the sample estimates indicate they do?

6.4 THE SAMPLE SIZE CONDITION

Whether or not the result of the central limit theorem is valid is also dependent on the size of the sample used. This is known as the sample size condition. We will discuss this condition for each of the four population parameters of interest.

1. Population Mean

In this case, when deciding whether our sample size is large enough for the central limit theorem to be valid, we should start by constructing a histogram of our individual sample measurements. As we discussed in Chapter 4, the histogram (if based on a random sample of individual measurements) should be a good reflection of the approximate shape of the population distribution. If we determine from the histogram that the population distribution of individual values is approximately normal, then according to the central limit theorem, the distribution of possible sample means will follow the normal distribution, no matter what sample size is used.

If you determine from the histogram that the population distribution of individual values is skewed, our sample size will need to be greater than or equal to 40 to ensure that the distribution of possible sample means follows the normal distribution.

The sample size of 40 is a rule of thumb. In some cases, where the population distribution of individual measurements is extremely non-normal or skewed (like household incomes), then the sample size may need to be somewhat greater than 40 for the distribution of possible sample means to follow the normal distribution. However, a sample size of at least 40 should ensure the central limit theorem is valid in most cases.

2. Population Mean Difference

When estimating a population mean difference for quantitative data, the sample size condition (described above for a single population mean) applies to both groups. A histogram of individual measurements for both groups should be constructed. If both histograms approximately follow a normal distribution, then the distribution of possible sample mean differences will follow the normal distribution, regardless of the sample size in each group.

If the histogram is non-normal or skewed for one or both groups, then the sample size for the group(s) needs to be greater than 40 for the result of the central limit theorem to be valid. If either sample size is less than 40, and the corresponding histogram of individual measurements is non-normal or skewed, then the distribution of possible sample mean differences will be non-normal or skewed. The assumption that the distribution of sample mean differences is normal centered at the population mean difference would not be valid.

3. Population Proportion

In this case, there are two conditions that need to be checked to ensure a sufficient sample size for the central limit theorem to be valid. These conditions will differ slightly, depending on whether we are making statistical decisions using hypothesis testing or **confidence intervals**, both discussed in the following chapters.

With hypothesis testing (discussed in Chapter 8), we begin with a value for the population proportion in the **null hypothesis**, known as the **null value**. It is the currently accepted or assumed value for the population proportion.

The first condition is that the sample size multiplied by the null value for the population proportion must be greater than or equal to 10. The second condition is that the sample size multiplied by the one minus the null value needs to be greater than or equal to 10. If both conditions are met, the sample sizes are large enough for the result of the central limit theorem to be valid, and therefore the results of our hypothesis test will be valid.

When constructing confidence intervals (discussed in Chapter 7), we make no assumptions about the value of the population proportion. Our sample proportion is an estimate of this value.

In this case, the first condition is that the sample size multiplied by the sample proportion needs to be greater than or equal to 10. The second condition is that the sample size multiplied by the one minus the sample proportion must be greater than or equal to 10. If both conditions are met, the sample size is large enough for the result of the central limit theorem to be valid, and therefore our confidence interval will be valid.

4. Population Proportion Difference

When estimating a population proportion difference, the two conditions described above (for a single population proportion) must be checked for both groups. If both conditions are met, the sample sizes for both groups are large enough for the result of the central limit theorem to be valid.

If the sample size conditions are satisfied, it does not necessarily mean that our sample statistic will be close to the population parameter of interest. It should be remembered that the larger the sample size, the smaller the standard error, the closer we expect our sample statistic to be (on average) to the value of the population parameter. A particular sample size is required for the distribution of possible sample statistics to follow the normal distribution, but a larger sample size will improve accuracy in our estimation of the population parameter.

6.5 PREDICTING ELECTION RESULTS

From a statistical point of view, elections are interesting because it is one of the rare situations where we will know the proportion of the voters (the population parameter) who voted for a particular candidate, once the results are counted.

In the run-up to a national election, particularly in the United States, there will be numerous organizations conducting election polling. From what we have learned in this chapter, we can say that we expect the average of the results of all polls conducted to be close to the result of the election. According to the central limit theorem, the average of all possible sample proportions is the population proportion. The more polls that are conducted, the closer we expect the average of the polls to be to the true outcome of the election. For this to be true, we have to assume that each pollster used a random sample of voters, valid weighting was implemented when necessary, and the same sample size was used for each poll. In reality, the polls would not be based on the same sample size.

Different sample sizes are not a big issue when calculating the average of poll results. As long as each poll was otherwise well conducted, the average of poll results can be calculated by weighting each individual poll result (according to sample size used) in the calculation. The website Real Clear Politics calculates the average of polls leading up to general elections to try and predict the election result.

Table 6.1: General Election 2008: McCain versus Obama

Poll	Date	Sample	MoE	Obama (D)	McCain(R)	Spread
Final Results	–	–	–	52.9	45.6	Obama +7.3
RCP Average	10/29–11/3	–	–	52.1	44.5	Obama +7.6
Marist	11/3–11/3	804	4.0	52	43	Obama +9
Battleground (Lake)	11/2–11/3	800	3.5	52	47	Obama +5
Battleground (Tarrance)	11/2–11/3	800	3.5	50	48	Obama +2
Rasmussen Reports	11/1–11/3	3000	2.0	52	46	Obama +6
Reuters/C-SPAN/Zogby	11/1–11/3	1201	2.9	54	43	Obama +11
IBD/TIPP	11/1–11/3	981	3.2	52	44	Obama +8
Fox News	11/1–11/2	971	3.0	50	43	Obama +7
NBC News/Wall St. Journal	11/1–11/2	1011	3.1	51	43	Obama +8

Poll	Date	Sample	MoE	Obama (D)	McCain(R)	Spread
Gallup	10/31–11/2	2472	2.0	55	44	Obama +11
Diageo/Hotline	10/31–11/2	887	3.3	50	45	Obama +5
CBS News	10/31–11/2	714	–	51	42	Obama +9
Ipos/McClatchy	10/30–11/2	760	3.6	53	46	Obama +7
ABC News/ Wash Post	10/30–11/2	2470	2.5	53	44	Obama +9
CNN/Opinion Search	10/30–11/1	714	3.5	53	46	Obama +7
Pew Research	10/29–11/1	2587	2.0	52	46	Obama +6

Citation: http://www.realclearpolitics.com/epolls/2008/president/us/general_election_mccain_vs_obama-225.html

In the 2008 general election on November 4 between Barack Obama and John McCain, the election result was 52.9% for Obama and 45.6% for McCain. The average of the polls leading up to election (from October 29th to November 3rd) was 52.1% for Obama and 44.5% for McCain, not very far from the actual results.

A *Huffington Post* article titled "Journalists Should Stop Highlighting Individual Polls and Focus on Polling Averages" discusses the reasons why journalists should focus on polling averages and not individual polls. The first reason given is the one we have discussed in this chapter: the average result of many polls is much more likely to be closer to the truth than the results of an individual poll. The second reason given was as follows:

> People are predisposed to dismiss individual polls altogether when the results suggest that they hold minority views or when their preferred candidate is losing. That is not the case with polling averages. Individuals are far less likely to dismiss these results on partisan grounds.

The writers of the news article go on to discuss how they tried to test their theory in the lead-up to the 2016 US general election. With such a partisan election, they found that both Clinton and Trump supporters rejected individual polls and polling averages when they

GENERAL ELECTION 2008: McCAIN VERSUS OBAMA

Web Link: http://www.realclearpolitics.com/epolls/2008/president/us/general_election_mccain_vs_obama-225.html

Search Term: Real Clear Politics McCain versus Obama

NEWS ARTICLE: JOURNALISTS SHOULD STOP HIGHLIGHTING INDIVIDUAL POLLS AND FOCUS ON POLLING AVERAGES

Web Link: https://www.huffingtonpost.com/entry/journalists-polling-averages_us_59dbaa19e4b0b34afa5b37a7

Search Term: Journalists Should Stop Highlighting Individual Polls

pointed to the other candidate who was leading. However, when they broke down the electorate by education level, they found that partisan bias shrunk among educated voters when it came polling averages versus individual polls.

The lead-up to the 2016 US general election was probably not the best time to test the theory that people are more inclined to believe polling averages. Both candidates were so polarizing that most of the electorate did not want to believe that the other candidate could be elected president. However, it does make (intuitive) sense that people would put more weight on polling averages than individual polls. As the result of the central limit theorem shows: the average of all the possible sample statistics is the truth.

6.6 CONCLUSION

We have defined a framework for describing how sample statistics are distributed around an unknown population parameter. As long as we adhere to the necessary assumptions and conditions regarding the quality of our data and sample size, our sample statistic is one of a range of possible sample statistics that distribute themselves according to the normal distribution, centered at the unknown population parameter of interest. The larger the sample sizes, the narrower the range of possible sample statistics. We measure the variation in sample statistics around the population parameter using the standard error.

In the next chapter, we use this framework to construct confidence intervals for making statistical decisions about the four population parameters we discussed: population mean, population mean difference, population proportion, and population proportion difference.

6.7 REAL-WORLD EXERCISES

1. Toss a coin a 200 times. After every four tosses, record the number of heads you get. Create a histogram of the number of heads (0, 1, 2, 3, 4) you got in every four tosses. You can think of every four coin-tosses as your sample size in a repeated experiment. Compare the shape of your histogram to what you expect the shape to be, according to the central limit theorem.

2. Looking at the **2011–2012 NHANES** data in Chapter 4, we found that in a random sample of 2,742 men, the sample mean height of men was 174 cm. You can download the dataset from the NHANES website. It is an SAS dataset, so you will need SAS University Edition (free statistical software) to view it. Alternatively, find a sample dataset containing a quantitative variable with at least 1,000 observations. Calculate the sample mean.

Let's suppose the sample data can be thought of as the entire population of values. If so, complete the following:

a. Select a random sample of 16 men from the dataset and calculate the mean. You can randomly select 16 men from the dataset using the random number generator at https://www.randomizer.org/. Enter 1 to 2,742 (or the number of observations in the dataset you selected) as the range of numbers (observations) you want to select from. Generate 16 numbers in this range, representing the 16 individuals or observations you will select from the dataset.

b. Repeat the process described in a. 40 times. Create a histogram of the 40 sample means based on sample sizes of 16 men.

c. Comment on the shape of the histogram as it relates to the central limit theorem and the empirical rule for the normal distribution.

2011–2012 NATIONAL HEALTH AND NUTRITION EXAMINATION SURVEY (NHANES)

Web Link: https://wwwn.cdc.gov/Nchs/Nhanes/Search/DataPage.aspx?Component=Examination&CycleBeginYear=2011

Search Term: NHANES 2011–2012 Examination Data

3. Try to find a large dataset, preferably population or census data containing a quantitative variable. One possible source is www.data.gov. Download the dataset and complete the following:

a. Calculate mean and standard deviation for the quantitative variable based on all the observations. Treat the mean and standard deviation as population values.

b. Using a random number generator (www.randomizer.org/) or statistical software, select two random samples of size 9 and size 36 from the dataset. Calculate the mean and standard deviation for both samples.

c. Calculate the standard error of the sample means for both samples, and explain what they mean.

d. Using the standard errors, draw the normal distribution of possible sample means based on both sample sizes.

e. Explain why the spread of possible sample means around the population mean are different for each sample size.

4. Referring to Q3, using a random number generator (www.randomizer.org/) or statistical software:

   a. Select 40 random samples of size 9 and 40 random samples of size 36 and calculate sample means for each sample.

   b. Create histograms of sample means for both sample sizes.

   c. Comment on the shape and spread of the histograms as compared to the normal distribution of sample means drawn in part d) of Q3.

5. Try to find a large dataset, preferably population or census data containing a categorical variable. For example, a Yes/No variable. One possible source is www.data.gov. Download the dataset and complete the following:

   a. Calculate the proportion of Yes answers based on all the observations. Treat the proportion as if it were a population proportion.

   b. Using a random number generator (www.randomizer.org/) or statistical software, select two random samples of size 25 and size 100 from the dataset. Calculate the sample proportions for both samples.

   c. Calculate the standard error of the sample proportions for both samples and explain what they mean.

   d. Use the standard errors, and draw the normal distribution of possible sample proportions based on both sample sizes.

   e. Explain why the spread in possible sample proportions around the population proportion is different for each sample size.

6. Referring to Q5, using a random number generator (www.randomizer.org/) or statistical software:

   a. Select 40 random samples of size 25 and size 100, respectively, and calculate sample proportions for each sample.

   b. Create a histograms of sample proportions for both sample sizes.

   c. Comment on the shape and spread of the histograms as compared to the normal distribution of possible sample proportions drawn in part d) of Q5.

IMAGE CREDIT

Tbl 6.1: Source: http://www.realclearpolitics.com/epolls/2008/president/us/general_election_mccain_vs_obama-225.html.

MAKING STATISTICAL DECISIONS WITH CONFIDENCE INTERVALS

CHAPTER 7

In this world nothing can be said to be certain, except death and taxes.

—Benjamin Franklin

7.1 INTRODUCTION

In our last chapter, we learned how to reason with and describe the variation in sample statistics for four primary population parameters of interest: population mean, population mean difference, population proportion, and the population proportion difference. We learned that when certain assumptions and conditions are met, sample statistics (based on a particular sample size) follow the normal distribution with a mean equal to the population parameter of interest. We described the variation in sample statistics around the population parameter using the standard error and the empirical rule for the normal distribution.

The result of the central limit theorem provides us with a framework for making statistical decisions about population parameters of interest. In this chapter, we will construct and reason with what are known as **confidence intervals**; one approach to reasoning with this result to find answers to our statistical questions. We will begin to reason with the uncertainty due to sampling variation regarding the value of the four population parameters of interest by asking the following questions:

> **KEY TERMS**
>
> **Confidence Interval:** An interval of plausible values for the population parameter of interest
>
> **Confidence Level:** A measure of the degree of reliability in the confidence interval

- How can we construct an interval containing plausible values for the population parameter?
- Why is the interval we construct called a confidence interval?
- What do we mean when we say we are 95% confident that our confidence interval contains the population parameter?
- What do we mean by different confidence levels?

We will learn how to reason with the uncertainty regarding the value of the population parameter of interest by trying to capture the value within a confidence interval. We will reason with the likelihood that the confidence interval does in fact contain the true value of the population parameter. The confidence interval will be constructed using the same method for each of the four parameters of interest. However, we need to understand how to explain what the confidence interval means in terms of the particular population parameter of interest in question.

7.2 CONFIDENCE INTERVAL FOR THE POPULATION PARAMETER

The first method we will use to make statistical decisions regarding the value of the population parameter of interest is called a confidence interval. We can say with a certain degree of reliability, most commonly being 95%, that our population parameter is one of a range of plausible values within the interval. For this method to be valid, it is very important that the sample data collected adheres to the necessary assumptions and conditions discussed in the last chapter. We are assuming our sample statistic is one of a range of possible sample statistics distributed under the normal distribution, centered at the unknown population parameter.

The fact that the population parameter is unknown makes it impossible to know how far exactly our sample statistic deviates from the population parameter. However, we do know that (given the necessary assumptions and conditions are met) approximately 95% of possible sample statistics will be within two standard errors of the population parameter. We will use this knowledge to construct our confidence interval for the population parameter.

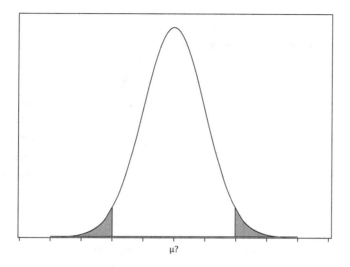

Figure 7.1: Unknown Population Parameter and Sampling Distribution

Data source: http://blogs.sas.com/content/sastraining/2014/06/10/producing-normal-density-plots-with-shading/

We will use a simple analogy to help us understand what confidence intervals are and how they are constructed. Imagine that the (measurement) line drawn below the normal distribution in Figure 7.1 is a long and straight stretch of the Mississippi River. The normal distribution drawn above the line is a mountain not far from the riverside. One particular day, a fisherman in his boat (the sample statistic) is riding along this stretch of the river hoping to capture what is known as the great fish (the population parameter) of the Mississippi. At this particular time of year, the great fish is known to be sleeping in the part of the river directly beneath the mountain peak. For the purpose of the analogy, let's assume that fish sleep!

The problem the fisherman has is that with large trees covering the riverbank, he can't see the mountain and thus his exact location in relation to the great fish. However, he knows his nets can extend two standard errors in both directions from his location. Therefore, he is highly confident that he will catch the fish within his nets because there is a 95% probability that his boat is within two standard errors of where the fish lies sleeping.

So what he decides to do is cast his nets two standard errors in both directions, up and down the river. He knows that if his boat is within two standard errors of the fish, (between the gray shaded areas in Fig. 7.1), then the fish will get captured somewhere within his nets. If his boat is more than two standard errors away from the fish (in the gray shaded areas in Figure 7.1), he won't capture the fish within his nets. There is a 5% probability of this happening.

By fishing in this way, the fisherman can be 95% confident that he will catch the fish within his nets. The 95% (or 0.95) is a probability value known as the **confidence level**. From our point of view, we know that (if this process could be repeated over and over) the fisherman will be within two standard errors of the great fish 95% of the time, meaning that 95% of the time he would capture the fish somewhere within his nets.

In other words, when we select a random sample of a particular size, we know (from the result of the central limit theorem) that our sample statistic is one of a range of possible sample statistics normally distributed around the population parameter. Consequently, there is a 95% probability that our sample statistic is within two standard errors of the population parameter. By adding and subtracting two standard errors to and from our sample statistic, we can calculate upper and lower limits for an interval of plausible values for the population parameter. As a result, we can say that we are 95% confident that the population parameter is within that interval.

The general form of a 95% confidence interval is as follows:

$$\text{sample statistic} \pm 2 \times (\text{standard error})$$

The 95% probability (our confidence level) refers to the confidence we have in the process. Imagine selecting numerous random samples of the same size and constructing a confidence interval based on each sample. As you increase the number of times you repeat this process, the percentage of the confidence intervals containing the population parameter will approach 95%.

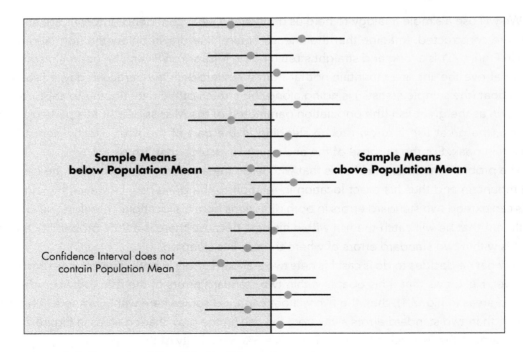

Figure 7.2: Example of Possible Confidence Intervals

Figure 7.2 illustrates this point by constructing twenty confidence intervals based on the same sample size. The blue dots represent the sample means, and the horizontal black lines represent the width of the confidence intervals. The vertical black line represents the population mean. The upper and lower limits of the confidence intervals will differ from sample to sample, and we expect nineteen out of every twenty 95% confidence intervals to contain the population parameter.

The range of plausible values within the confidence interval can be thought of as a probability map. A good comparative example is the probability map used for predicting the path of hurricanes. In September 2017, the center of the probability map projecting the path for Hurricane Irma suggested that Miami, Florida would get a direct hit. However, as the article in the *New York Times* titled "*Irma Shifting Forecasts: It's All a Matter of Probability*" pointed out:

> Many people trying to use forecasts like those provided by the National Hurricane Center, however, do not fully understand the cone of probability and focus instead on the line that runs down the middle, taking it as an accurate prediction of the storm's path.
>
> J. Marshall Shepherd, an atmospheric scientist at the University of Georgia, explained the fallacy in a Facebook post. "Anywhere in that cone is a possibility," Dr. Shepherd wrote, "and it has always been a challenge communicating what the cone 'means' versus what people 'think' it means."

The line that runs through the middle of the probability map (or cone of probability) for Hurricane Irma is analogous to our sample mean in the middle of our confidence interval. It turned out that the center of the storm went somewhat west of Miami, hitting the Florida Keys, but was within the area covered by the probability map. This is analogous to our population mean being somewhat to the left of our sample mean within our confidence interval.

It would be interesting to know what was the standard error in this example and how many standard errors Florida Keys was from Miami. In Chapter 6, we learned that the distribution of sample means follows the normal distribution. Therefore, according to the empirical rule for the normal distribution, we can say 68% of sample means (based on a particular sample size) should be within one standard error of the population mean. In other words, more often than not, our sample statistic will be no more than one standard error from the population parameter. We can say 95% of sample means should be within two standard errors of the population mean. In other words, it is highly probable our sample statistic will be no more than two standard errors from the population parameter. With a large sample of quality data resulting in a small standard error, it is highly probable that our sample statistic will be very close to the population parameter even if it is two standard errors away. This is the power of quality data and statistics as a means of pursuing truth.

IRMA SHIFTING FORECASTS: IT'S ALL A MATTER OF PROBABILITY

Web Link: https://www.nytimes.com/2017/09/10/us/forecast-irma-shift-florida.html?_r=0

Search Term: Irma Shifting Forecasts

There is a subtle difference between the 95% (or 0.95) probability we are discussing here and the 95% "confidence level" probability for describing a confidence interval. For a confidence interval, the 95% probability is the probability the confidence interval we construct contains the population parameter. As already stated, if we were to repeat the process (of constructing a confidence interval with repeated random samples of the same size), we would expect 95 out of every 100 confidence intervals to contain the population parameter. This can feel confusing when you are first trying to gain a conceptual understanding of confidence intervals and what they mean. The confidence interval either contains the population parameter or it does not. Ninety-five percent of all 95% confidence intervals (based on a particular sample size) will contain the population parameter and 5% of them will not.

One final discussion regarding our conceptual understanding of confidence intervals relates to the range of possible values for the population parameter within the confidence interval itself. Is one value more likely to be the population parameter than another? The answer to this question is that the most likely value for the population parameter is our sample statistic, the midpoint of our confidence interval, given the data we collected. Sir Ronald Fisher, considered the father of modern statistics, developed what he termed likelihood functions that enable us to derive maximum likelihood estimates for the population parameter. A deeper discussion of these concepts is beyond the scope of this book. All you need to know is that for the different population parameters we pursue, the statistics we calculate from our sample data are considered the most likely values (or maximum likelihood estimates) for the population parameters of interest.

However, it is important to understand that the values near the midpoint of the confidence interval are more likely to be the population parameter than the values closer to the lower and upper limits of the confidence interval. When making statistical decisions using confidence intervals, we need to allow for the fact that the population parameter is likely to be anywhere within its limits or even just outside them. It should always be remembered that the accuracy of our confidence interval also depends on the quality of our data including our assumptions for valid analysis.

7.3 CONFIDENCE LEVELS

The most commonly used confidence level is 95%. However, there are others (90%, 99%) that we can use. Our intuition tells us that the more confident we want to be in our interval containing the population parameter, the wider the confidence interval will need to be. When constructing a 95% confidence interval, we multiplied the standard error by 2, which is a z-score (technically, we should be using a t-score for quantitative data). When constructing a 99% confidence interval, we multiply the standard error by 2.58, resulting in a wider confidence interval. When constructing a 90% confidence interval, we multiply the standard error by 1.65, resulting in a narrower confidence interval.

If we want greater confidence that our interval contains the population parameter, the trade-off is that we have to reason with a wider range of plausible values for the population parameter. In other words, greater confidence will result in less precision.

As already stated, the confidence level refers to the confidence we have in the process. If we were to select a very large number of samples of the same size and calculate a 95% confidence interval each time, we would expect 95% of the confidence intervals to contain the population parameter; whereas 5% of the confidence intervals would not.

7.4 CONSTRUCTING CONFIDENCE INTERVALS

In this section, we will construct confidence intervals for each of the four population parameters of interest: population mean, population mean difference, population proportion, and the population proportion difference.

7.4.1 Population Mean

We will apply the general form of the confidence interval to construct a 95% confidence interval for the population mean.

Example 7.1

In the (hypothetical) **Example 6.1**, we constructed the sampling distribution of sample mean heights of men in the United States. The sampling distribution was based on sample sizes of 36 men and assuming the population mean height is equal to 70 inches. Let's say we want to question that assumption. We decide to collect our own sample data and construct a 95% confidence interval for the population mean height.

We select a random sample of 36 men, resulting in a sample mean height equal to 69.5 inches and a sample standard deviation of 3 inches. The standard error is equal to 0.5 inches (see appendix for calculation). Our 95% confidence interval for the population mean height is:

$$\boxed{\bar{x} \pm t\left(\frac{s}{\sqrt{n}}\right)}$$

sample mean ± 2 x (standard error)
69.5 ± 2 x (0.5)
69.5 ± 1
[68.5, 70.5]

We can say with 95% confidence that the population mean height of men lies between 68.5 to 70.5 inches. Therefore, based on the data we collected, it is quite plausible that the population mean height of men is equal to 70 inches.

According to the central limit theorem, for the confidence interval to be valid, we must remember to check the necessary assumptions and conditions. It is fair to assume that the height measurements are independent of each other. The sample size is less than 40, so we have to assume that the population distribution of individual height measurements is normally distributed. From the height data in the NHANES study we looked at in Chapter 4, we know that this is a fair assumption to make.

If we were to increase the sample size to 144 men, obtain the same sample mean height of 69.5 inches and sample standard deviation of 3 inches, the standard would be equal to 0.25 inches (see appendix for calculation). Our 95% confidence interval for the population mean height is as follows:

$$69.5 \pm 2 \times (0.25)$$
$$69.5 \pm 0.5$$
$$[69, 70]$$

We can now say with 95% confidence that the population mean height of men lies between 69 and 70 inches, a narrower range of plausible values for the population mean.

Obviously, the sample mean height of 69.5 inches, based on a sample of 144 men, is just as close to the unknown population mean as the same sample mean height based on sample size of 36 men. From the central limit theorem, we learned that a larger sample size will result in a narrower range of possible sample means around the unknown population mean.

Therefore, a larger sample size (resulting in a smaller standard error) enables us to make a more precise statement about how close the sample mean is to the population mean than we could with a smaller sample size. As we discussed in section 7.2, there is a 95% probability that the sample mean is no more than two standard errors from the population mean. In this example, this means it is highly probable that the distance between the sample mean and population mean is no more than 0.5 inches. That is quite a small distance, especially since our sample mean is based on a sample size of only 144 men from a much larger population.

Even though this is a hypothetical example, the standard errors, based on each sample size, are pretty close to what you would expect for men's height data. The confidence intervals are very narrow even though our sample sizes are relatively small. This is the power of proper random sampling to give us accurate estimates of population parameters. The size of the population does not determine how close our sample statistic is to the population parameter. The accuracy of our statistics is driven by the sample size and measured using the standard error. In a well-conducted study, where a proper random sample is selected and accurate measurements are taken, a relatively small sample size can produce accurate estimates of population parameters, especially when natural variation in measurements is not too large (like the variation in men's heights). The smaller the standard error, the closer we expect our estimates to be to the truth.

Example 7.2

In Chapter 4, we looked at a sample of 2,742 men's weights from the 2011–2012 NHANES study. The sample mean weight was 85 kg with a sample standard deviation of 20.6 kg. The standard error is equal to 0.39 kg (see appendix for calculation).

The resulting 95% confidence interval for the population mean weight is as follows:

$$85 \pm 2 \times 0.39$$
$$[84.22, 85.78]$$

The large sample size (resulting in a very small standard error) enables us to make a very precise statement regarding the plausible values for the population mean weight of men in the United States. We can say with 95% confidence that the population mean weight of men in the United States lies between 84.22 kg and 85.78 kg, a very narrow range of plausible values.

The NHANES study also contained a sample of 2,794 women's weights. The sample mean weight was 75 kg with a sample standard deviation of 20.9 kg. The standard error is equal to 0.40 kg (see appendix for calculation).

The resulting 95% confidence interval for the population mean weight is as follows:

$$75 \pm 2 \times 0.40$$
$$[74.2, 75.8]$$

We can say with 95% confidence that the population mean weight of women in the United States lies between 74.2 kg and 75.8 kg.

The confidence interval for the population mean weight of women lies completely below the confidence interval for the population mean weight of men. Using these confidence intervals, we can state that we have found statistical evidence that the population mean weight of men is different from the population mean weight of women. In the next section, we will make a more direct comparison of mean weights by calculating a confidence interval for the population mean difference (in weight) between men and women.

Example 7.3: The Fast Diet Study

We will now take a look at the confidence intervals presented in the Fast Diet study we critiqued in Chapter 2. As a reminder, the subjects were randomized to two diets: The Intermittent Energy Restriction (IER) diet was where the subjects went on an extremely low-calorie diet for

> **FAST DIET STUDY FROM CHAPTER 2**
>
> The Effects of Intermittent or Continuous Energy Restriction on Weight Loss and Metabolic Disease Risk Markers: A Randomized Trial in Young Overweight Women
>
>
>
> Web Link: http://www.nature.com/ijo/journal/v35/n5/full/ijo2010171a.html
>
> Search Term: The Effects of Intermittent or Continuous Energy Restriction

two days a week, and a regular diet called the Continuous Energy Restriction diet (CER). According to the journal article:

> **Results:** *Last observation carried forward analysis showed that IER and CER are equally effective for weight loss: mean (95% confidence interval) weight change for IER was −6.4 (−7.9 to −4.8) kg vs −5.6 (−6.9 to −4.4) kg for CER*

The study provided two separate confidence intervals for population mean weight loss on both diets. It would have been more informative to have provided a confidence interval for the population mean difference in weight loss between the two diets, the actual parameter of interest in this study.

For these confidence intervals to be valid, we have to assume that the individual weight loss measurements were independent of each other. If some of the individuals were working together to lose weight, then the assumption of independence (of the measurements) would be called into question.

The sample sizes were 53 subjects and 54 subjects on the IER and CER diets, respectively. The distribution of individual weight loss measurements (for each group) may not be normally distributed. However, according to the central limit theorem, the sample sizes are large enough for the confidence intervals to be valid.

For the IER diet, the sample mean weight loss was 6.4 kg with a 95% confidence interval ranging from −7.9 kg to −4.8 kg. The researchers were 95% confident that the population mean weight loss on the IER diet would lie somewhere between 4.8 kg and 7.9 kg.

For the CER diet, the sample mean weight loss was 5.6 kg with a 95% confidence interval ranging from −6.9 kg to −4.4 kg. The researchers were 95% confident that the population mean weight loss on the IER diet would lie somewhere between 4.4 kg and 6.9 kg.

As mentioned, a confidence interval for the population mean difference in weight loss would have been more informative. If that confidence interval included 0 kg, it would suggest that it is possible there is no difference in the effectiveness of the two diets on average. However, by comparing the ranges of the two individual confidence intervals, we come to the same conclusion. There is a lot of overlap in the range of plausible values for the population mean weight loss in both confidence intervals, suggesting that the two diet programs are likely to be equally effective at reducing weight. Finally, we should remember the sample of women chosen for this study was far from representative of all women in the UK. Therefore, we should be cautious regarding what population we extend these results to.

No matter what you eat or what diet you are on, there is one effect from doing so that you can't avoid. That is flatulence, or what is more commonly known as farts! We all do it (though some of us like to pretend we don't), but it takes a child to ask the important question: How big is a fart? The FiveThirtyEight article titled "How Big Is a Fart" is a very amusing and educational article that tries to answer this question and other questions related to farts (or flatulence, if you prefer).

The following is an excerpt from the 2012 study mentioned in the article, describing the aims of the research.

Anal Gas Evacuation and Colonic Microbiota in Patients with Flatulence: Effect of Diet

Hence, our aims were to determine: (a) the effect of diet on the number of gas evacuations and the volume of gas evacuated; (b) whether the number of evacuations and the gas volume evacuated are increased in patients complaining of flatulence; and (c) the relationships among diet, abdominal sensations, gas evacuation and colonic microbiota.

The study sample size was 50 subjects consisting of "30 patients complaining of excessive passage of gas" and "20 healthy subjects without gastrointestinal symptoms". The measurement used was the "number of gas evacuations per anus using an event marker (DT2000 Memory Stopwatch, Digi sport instruments, Shanggiu, China) during each study phase."

The sample statistics for patients and healthy subjects were presented, means plus and minus the standard errors as follows:

Gas Evacuations by an Event Marker: Patients: 21.9 ± 2.8
Healthy Subjects: 7.4 ± 1.0

Using the sample means and standard errors, we can construct a 95% confidence interval for population mean number of gas evacuations per daytime for both the patients and the healthy subjects. For our confidence intervals to be valid, we have to assume that the fart measurements (number of farts) across individuals were independent of each other. This is a fair assumption to make! The sample sizes are small for both groups, so the histograms of the

HOW BIG IS A FART? SOMEWHERE BETWEEN A BOTTLE OF NAIL POLISH AND A CAN OF SODA

Q: How much space does a fart take in your body? — Inbal R., age 5

Placed under the microscope, even the dullest grain of sand develops a personality. So it goes with farts. (Or "flatulence," as they say in the scientific literature.) Farts may seem largely interchangeable, but each one is special. Even just your own farts are a circus sideshow of intestinal gas: big ones, little ones, stinky ones, oddly fresh ones. There is not enough scientific evidence to say that no two farts are alike — but you can rest assured they are a riot of diversity.

....

In 2012, for instance, researchers took healthy volunteers and those who suffered from chronic gastrointestinal problems, fed them either a neutral or fart-inducing breakfast, and then put a catheter up each of their anuses to collect farts and transfer the gas to a machine that measured the volume of those farts in real time.

Web Link: http://fivethirtyeight.com/features/how-big-is-a-fart-somewhere-between-a-bottle-of-nail-polish-and-a-can-of-soda/

Search Term: How Big Is a Fart?

ANAL GAS EVACUATION AND COLONIC MICROBIOTA

Web Link: http://gut.bmj.com/content/63/3/401.long

Search Term: Anal Gas Evacuation and Colonic Microbiota

individual measurements should be (approximately) normally distributed to satisfy the sample size condition. Though not discussed in the journal article, we have to assume the researchers checked the sample size condition before completing their analysis.

For the patients, the 95% confidence interval is calculated as follows:

$$21.9 \pm 2 \times 2.8$$
$$21.9 \pm 5.6$$
$$[16.3, 27.5]$$

The interval tells us that we can be 95% confident that the mean number of daily farts in the population of patients complaining of excessive passing of gas lies between 16.3 and 27.5.

For the healthy subjects, the 95% confidence interval is calculated as follows:

$$7.4 \pm 2 \times 1.0$$
$$7.4 \pm 2$$
$$[5.4, 9.4]$$

The interval tells us that we can be 95% confident that the mean number of daily farts in the population of healthy subjects lies between 5.4 and 9.4. One conclusion we can take from this study is that we can be highly confident that everyone farts on average at least five times a day. Yes, that includes *you*!

7.4.2 Population Mean Difference

We will apply the general form of the confidence interval to construct a 95% confidence interval for the population mean difference.

Example 7.4

In the (hypothetical) **Example 6.2**, we constructed the sampling distribution for sample mean differences in heights between men and women in the United States. The sampling distribution was based on a total sample size of 93 men and women and assuming the population mean difference in height is equal to 3 inches. We selected a sample of data to estimate the population mean difference for ourselves.

We selected a random sample of 36 men and 57 women, with a sample standard deviation of 3 inches for men and 2.5 inches for women respectively. The resulting standard error for the sample mean difference was 0.6 inches (see appendix for calculation). The sample mean height for men was 69 inches and for women it was 65 inches, resulting in a sample mean difference in height (men minus women) equal to 4 inches. We can use this sample data to construct a confidence interval containing a range of plausible values for the population mean difference in height.

Our 95% confidence interval for the population mean difference in height is:

$$(\bar{x}_1 - \bar{x}_2) \pm t\,(se(\bar{x}_1 - \bar{x}_2))$$

sample mean difference ± 2 x standard error of difference
4 ± 2 x (0.6)
4 ± 1.2
[2.8, 5.2]

We can say with 95% confidence that the population mean difference in height between men and women lies between 2.8 and 5.2 inches. Since the confidence interval contains 3 inches, it is possible that the population mean difference in height is equal to this value.

Example 7.5

In **Example 7.2**, using the NHANES weight data, we calculated separate confidence intervals for the population mean weight of men and for the population mean weight of women. We will now calculate a confidence interval for the population mean difference in weight between men and women.

Table 7.1

Sample Groups	Male	Female
Sample Size	2,742	2,794
Sample Mean	85	75
Sample Standard Deviation	20.6	20.9

From Table 7.1, we calculate the sample mean difference in weight (men minus women) to be 10 kg. The standard error of the sample mean difference is equal to 0.56 kg (see appendix for calculation). A confidence interval for the population mean difference in weight will give us a range of plausible values for that difference.

Our 95% confidence interval for the population mean difference in weight is:

(85 − 75) ± 2 x 0.56
10 ± 1.12
[8.88, 11.12]

We can say with 95% confidence that the population mean difference in weight between men and women lies between 8.8 kg and 11.12 kg. The very large sample size for both men and women enables us to make a very precise statement about the difference in mean weight between men and women in the US population. There is little to no doubt that men are heavier than women on average, and given this sample data, we can be highly confident that the population mean difference in weight is between 8.88 kg and 11.12 kg.

We will now return to the study titled "Relationship of Collegiate Football Experience and Concussion with Hippocampal Volume and Cognitive Outcomes," discussed in Chapter 2:

> **OBJECTIVE:** *To assess the relationships of concussion history and years of football experience with hippocampal volume and cognitive performance in collegiate football athletes.*

The study sample consisted of 25 college football players with a history of concussion, 25 players with no history of concussion, and 25 healthy non-football players (controls) matched by age, sex and education. The researchers looked at the relationship of concussion to hippocampal volume and other cognitive outcomes. The journal states that:

> *The hippocampus is a brain region involved in regulating multiple cognitive and emotional processes affected by concussion and is particularly sensitive to moderate and severe traumatic brain injury.*

STUDY FIRST LOOKED AT IN CHAPTER 2: RELATIONSHIP OF COLLEGIATE FOOTBALL EXPERIENCE AND CONCUSSION WITH HIPPOCAMPAL VOLUME AND COGNITIVE OUTCOMES

Web Link: http://jamanetwork.com/journals/jama/fullarticle/1869211

Search Term: Relationship of Collegiate Football Experience and Concussion

The researchers found a difference in mean hippocampal volumes comparing players with and without concussion with the healthy controls. They also found a difference in mean hippocampal volumes between players with and without concussions. The sample mean differences were as follows:

- 1,788 uL: Controls minus Players with Concussion
- 1,027 uL: Controls minus Players without Concussion
- 761 uL: Players without Concussion minus Players with Concussion

We will look at the confidence intervals presented for the population mean differences in hippocampal volumes among the three groups. For the confidence intervals to be valid, we have to assume the individual measurements (for hippocampal volume) are independent within each group and across each group. It is fair to assume that hippocampal volume values would be independent from player to player.

The sample sizes are small in each group, so the histograms of the individual measurements for all three groups should be (approximately) normally distributed for the confidence intervals to be valid. Though not discussed in the journal article, we will assume the researchers checked the sample size condition.

In this example, the journal article did not provide the standard errors for the sample mean difference. Only the sample mean differences and resulting confidence intervals were presented.

Controls versus Players with Concussion

The sample mean difference (controls minus players with concussion) in hippocampal volume was 1788 uL. The 95% confidence interval for the population mean difference between these two groups was [1,317, 2,258]. When comparing controls with the players with concussion, we are 95% confident that the population mean difference in hippocampal volume lies between 1,317 uL and 2,258 uL.

Controls versus Players without Concussion

The sample mean difference (controls minus players without concussion) in hippocampal volume was 1,027 uL. The 95% confidence interval for the population mean difference between these two groups was [556, 1,498]. When comparing controls with the players without concussion, we are 95% confident that the population mean difference in hippocampal volume lies between 556 uL and 1,498 uL.

Players without Concussion versus Players with Concussion

The sample mean difference (players without concussion minus players with concussion) in hippocampal volume was 761 uL. The 95% confidence interval for the population mean difference between these two groups was [280, 1,242]. When comparing players without concussion with players with concussions, we are 95% confident that the population mean difference in hippocampal volume lies between 280 uL and 1,242 uL.

The results of the analysis show strong statistical evidence of a relationship between football playing and hippocampal volume. Players with concussion have (on average) significantly lower levels of hippocampal volume compared to the players without concussion and the controls. Players without concussion have (on average) significantly lower levels of hippocampal volume compared to the controls. Since the study was based on observational data, we can't make causal conclusions. However, the large differences in hippocampal volume across groups (though based on small samples) should be cause for concern and further study.

7.4.3 Population Proportion

We will apply the general form of the confidence interval to construct a 95% confidence interval for the population proportion.

Example 7.6

Let's say that we are interested in estimating the population proportion of people who are going to vote for a particular candidate in an election. We select a random sample of 500 voters and ask them who they are going to vote for. We find that 200 out of the 500 voters plan to vote for the candidate, a sample proportion of 200/500 equal to 0.40. The standard error is equal to 0.02 (see appendix for calculation).

Our 95% confidence interval for the population proportion is:

$$\hat{p} \pm z(se(\hat{p}))$$

sample proportion ± 2 x standard error of sample proportion

0.4 ± 2 x (0.02)

0.4 ± 0.04

[0.36, 0.44]

We can say with 95% confidence that the population proportion of voters who will vote for the particular candidate in the election lies between 0.36 and 0.44, or between 36% and 44%.

In Chapter 3, we discussed the results of a survey conducted by the Pew Research Institute titled "Political Polarization and Media Habits." The institute completed an extensive survey on the polarization of the American electorate and how both conservatives and liberals get their news on politics:

> When it comes to getting news about politics and government, liberals and conservatives inhabit different worlds. There is little overlap in the news sources they turn to and trust. And whether discussing politics online or with friends, they are more likely than others to interact with like-minded individuals, according to a new Pew Research Center study.

STUDY FIRST LOOKED AT IN CHAPTER 3: POLITICAL POLARIZATION & MEDIA HABITS

Web Link: https://www.journalism.org/2014/10/21/political-polarization-media-habits/political-polarization-and-media-habits-final-report-7-27-15/

Search Term: Political Polarization & Media Habits Final Report

There were many interesting findings in the report. One of the main findings was that 47% (or 0.47) of Americans whom the report defines as "consistently conservative" named Fox News as their main source for news on politics. The sample size was equal to 309 consistent conservatives with a margin of error of 7.2 percent (or 0.072).

Before calculating a 95% confidence interval for the population proportion (or percentage), we should point out the relationship between the margin of error (first discussed in Chapter 3) and the standard error. The relationship is as follows:

margin of error = 2 x standard error

For this example, if we calculate the margin of error (see formula given in Chapter 3) based on the sample size of 309, we find it to be equal to 5.7 percent (or 0.057). This value is smaller than the 7.2 percent (or 0.072) given in the report. The larger margin of error was a result of the necessary weighting due to an overall response rate of 61% in this study.

For the confidence interval to be valid, we have to assume the individual responses are independent of each other and that the sample size is sufficiently large. The study used a random sample, so we can assume the responses were independent. The sample size of 309 consistent conservatives is large enough to satisfy the sample size condition for proportions discussed in the last chapter.

The 95% confidence interval is calculated as follows:

$$\text{sample proportion} \pm \text{margin of error}$$
$$0.47 \pm 0.072$$
$$[0.398, 0.542]$$

We can say with 95% confidence that the population proportion of consistent conservatives who name Fox News as their main source for news on politics lies between 0.398 and 0.542, or between 39.8% and 54.2%. We will return to this survey in the next section when comparing proportions for two groups: consistent conservatives versus consistent liberals.

7.4.4 Population Proportion Difference

We will apply the general form of the confidence interval to construct a 95% confidence interval for the population proportion difference.

Example 7.7

In the (hypothetical) **Example 6.4**, we constructed the sampling distribution for sample proportion differences between men and women who vote for a particular candidate. The sampling distribution was based on a total sample size of 200 men and women and assuming the population proportion difference is equal to 0.20 (meaning 20% more men than women will vote for the candidate).

We selected a sample of 100 men and 100 women. We found that 55 of the men said they would vote for the particular candidate, a sample proportion equal to 0.55. We found that 45 of women said they would vote for the particular candidate, a sample proportion equal to 0.45. The sample proportion difference is 0.10, and the standard error is 0.07 (see appendix for calculation).

Our 95% confidence interval for the population proportion difference is:

$$\boxed{(\hat{p}_M - \hat{p}_F) \pm z(se(\hat{p}_M - \hat{p}_F))}$$

sample proportion difference ± 2 × standard error of difference
0.10 ± 2 × (0.07)
0.10 ± 0.14
[−0.04, 0.24]

We can say with 95% confidence that the population proportion difference lies between −0.04 and 0.24. This is quite a wide confidence interval. The inclusion of negative proportions in the interval means that it is possible that a higher proportion of women than men (in the population) will vote for the particular candidate. Since the confidence interval contains 0.20, it is possible that the population proportion difference is equal to this value.

Increasing the sample size to 400 for both men and women (and assuming we obtain the same sample proportions for men and women), the standard error is now equal to 0.035 (see appendix for calculation).

Our 95% confidence interval for the population proportion difference is:

0.10 ± 2 × (0.035)
0.10 ± 0.07
[0.03, 0.17]

We can say with 95% confidence that the population proportion difference lies between 0.03 and 0.17. The larger sample size for both groups has increased the precision with regard to what is the true value of the population proportion difference may be. Since the confidence interval does not contain 0.20, the data strongly suggests that the population proportion difference is equal to this value.

Let's return to the survey from the Pew Research Institute discussed in the last section. When it comes to absorbing political news through social media, two more interesting findings were as follows:

1. Consistent conservatives (CC) see more Facebook posts in line with their views than consistent liberals (CL), 47% versus 32%, respectively.

2. Consistent liberals (CL) are more likely to block others because of politics than consistent conservatives (CC), 44% versus 32%, respectively.

We will construct a confidence interval for the population proportion difference for both these questions. The standard error is equal to 0.044 (see appendix for calculation).

We can use this standard error in the calculation of the confidence intervals (for the population proportion difference), for both questions we are looking at. For the confidence intervals to be valid, we have to assume the individual responses are independent of each other, within each group and across each group, and that the sample sizes are sufficiently large. Again, since the study used a random sample, we can assume independence. The sample sizes of 309 consistent conservatives and 644 consistent liberals are large enough to satisfy the sample size condition for proportions discussed in the last chapter.

1. Our 95% confidence interval for the population proportion difference (CC minus CL) is:

$$(\hat{p}_{CC} - \hat{p}_{CL}) \pm z(se(\hat{p}_{CC} - \hat{p}_{CL}))$$

$(0.47 - 0.32) \pm 2 \times 0.044$
0.15 ± 0.088
$[0.062, 0.238]$

We can say with 95% confidence that the population proportion difference lies between 0.062 and 0.238, or between 6.2% and 23.8%. In other words, we can say that we are 95% confident that the percentage of consistent conservatives in the population who see Facebook posts in line with their views is somewhere between 6.2% and 23.8% greater than the percentage of consistent liberals in the population who see Facebook posts in line with their views.

2. Our 95% confidence interval for the population proportion difference (CL minus CC) is:

$(0.44 - 0.32) \pm 2 \times 0.044$
0.12 ± 0.088
$[0.032, 0.208]$

We can say with 95% confidence that the population proportion difference lies between 0.032 and 0.208, or between 3.2% and 20.8%. In other words, we can say that we are 95% confident that the percentage of consistent liberals in the population blocking others because of politics is somewhere between 3.2% and 20.8% greater than the percentage of consistent conservatives in the population who block others because of politics.

7.5 CONCLUSION

We have used the framework (the sampling distribution) we learned in Chapter 6, the result of the central limit theorem, to construct confidence intervals for our four population parameters of interest: population mean, population proportion, population mean difference, population proportion difference. The confidence interval provides us with a range of plausible values for the population parameter of interest.

We have learned that the larger the sample size, the smaller the standard error, the narrower the confidence interval. We have learned that the confidence level we use is the confidence we have in the process. The higher the confidence level, the wider the range of plausible values, and vice versa. In our next chapter, we will make use of this framework again to make statistical decisions regarding the value of population parameters using hypothesis testing.

7.6 REAL-WORLD EXERCISES

1. Exercise 2 in Chapter 3 asks you to design a survey asking ten questions, both categorical and quantitative. Use the data collected to complete the following:

    a. For a single categorical question, construct a 95% confidence interval for the population proportion.

    b. For a single quantitative question, construct a 95% confidence interval for the population mean.

    c. Compare two groups (e.g., Males and Females) on their response to a categorical question by constructing a 95% confidence interval for the population proportion difference.

    d. Compare two groups (e.g., Males and Females) on their response to a quantitative question by constructing a 95% confidence interval for the population mean difference.

    Before constructing each confidence interval, be sure to check the necessary assumptions and conditions for statistical analysis. State clearly what each of the confidence intervals mean in terms of the original question of interest.

MAKING STATISTICAL DECISIONS WITH CONFIDENCE INTERVALS | 133

2. Choose a place in your town or on your campus where you know many people will be passing by. Count the next 100 people or students passing, and note whether each person is doing one of the following: looking at his/her phone, wearing headphones, or carrying a backpack.

   a. Calculate a 95% confidence interval for the population proportion of people or students looking at their phones, wearing headphones, or carrying backpacks. Explain what the confidence interval means in context. Be sure to discuss and check whether your data adheres to the necessary assumptions and conditions for valid analysis.

   b. How representative do you think your sample is of the population of people in your town or the students on your campus? Briefly discuss.

3. Go to a polling website like https://www.realclearpolitics.com/. Choose the results of a Yes/No poll of interest to you that includes the margin of error or the sample size used.

   a. Using the margin of error or sample size used, calculate a 95% confidence interval for population proportion of interest. Explain what the confidence interval means in context. Be sure to discuss and check whether your data adheres to the necessary assumptions and conditions for valid analysis.

   b. Based on whatever information was given on how the sample data was collected, discuss how representative you think the sample is of the population the results were extended to.

4. Choose a place in your town or on your campus where you know many people will be passing by. For the next 36 males or females passing by, ask them one of the following: height, weight, how many hours he/she sleeps per night on average, or any other quantitative question of interest to you.

   a. Calculate a 95% confidence interval for population mean based on the quantitative variable type question you asked. Explain what the confidence interval means in context. Be sure to discuss and check whether your data adheres to the necessary assumptions and conditions for valid analysis.

   b. How representative do you think your sample is of the population of people in your town or the students on your campus?

5. Find data of interest to you online. From the dataset of your choosing, select a categorical variable and a quantitative variable.

   a. Calculate a 95% confidence interval for the population parameters of interest (population mean and population proportion) for both variables. Explain what the confidence intervals mean in context. Be sure to discuss and check whether your data adheres to the necessary assumptions and conditions for valid analysis.

   b. How representative do you think the data is of the population you are extending the results of your analysis to?

6. Find data of interest to you online. From the dataset of your choosing, select a quantitative variable as well as a categorical variable, for example, Gender. Group the values of the quantitative variable by the values of the categorical variable.

   a. Calculate a 95% confidence interval for the population mean difference across groups. Explain what the confidence intervals mean in context. Be sure to discuss and check whether your data adheres to the necessary assumptions and conditions for valid analysis.

   b. How representative do you think the data is of the population you are extending the results of your analysis to?

7. Collect data on a quantitative variable of interest to you where you can separate the variable values into two groups you wish to compare, for example, male and female.

   a. Calculate a 95% confidence interval for the population mean difference across groups. Explain what the confidence intervals mean in context. Be sure to discuss and check whether your data adheres to the necessary assumptions and conditions for valid analysis.

   b. How representative do you think the data is of the population you are extending the results of your analysis to?

CHAPTER 8

MAKING STATISTICAL DECISIONS WITH HYPOTHESIS TESTING

A man should look for what is, and not for what they think should be.

—Albert Einstein

8.1 INTRODUCTION

In our last chapter, we constructed and reasoned with confidence intervals as one approach to answering our statistical questions about our four primary population parameters of interest: population mean, population mean difference, population proportion, and the population proportion difference.

Another approach to making statistical decisions regarding the value of population parameters is known as **hypothesis testing**. There is one key difference between this approach compared to a confidence interval. With a confidence interval, we began with no assumptions regarding the value of the population parameter. We reasoned with our sample statistic and the sampling distribution centered at the unknown value for the population parameter and tried to capture that value within our confidence interval. With hypothesis testing, we begin with an assumed value for the population parameter, called the null hypothesized value, or simply the **null value**. We will reason with our sample statistic and the sampling distribution centered at an assumed (null) value for the population parameter of interest.

KEY TERMS

Hypothesis Testing: A method of statistical analysis used for making statistical decisions regarding the values of population parameters

Null Hypothesis: A general statement or default position regarding the value for a population parameter

Alternative (Research) Hypothesis: A statement rejecting the null hypothesis. When comparing treatments, it is a statement that the treatments are not equally effective. The alternative hypothesis is our research hypothesis

Null Value: The value for the population parameter presented in the null hypothesis

p-Value: The p-value is defined as the probability of getting the sample statistic we observed, or one more extreme (even further from the null value), given the null value is correct

Test Statistic: A standardized score used for calculating a p-value in hypothesis testing

Level of Significance: A borderline value (compared to the p-value) used to decide whether or not to reject the null hypothesis in favor of the alternative. The most commonly accepted borderline value is 0.05 or 5%. If the p-value is less than

0.05, the null hypothesis is rejected in favor of the alternative, and the results are considered statistically significant

Two-Sided Alternative: The alternative hypothesis used when it is believed that the population parameter could be less than or greater than the null value

One-Sided Alternative: The alternative hypothesis used when it is believed that the population parameter departs from the null value in one direction only

Cohen's d: A standardized score for measuring the magnitude of the sample effect size

Risk: a term used in epidemiology defined as the probability that an event (or outcome) will occur

Relative Risk: a value that compares one group's risk of developing a disease (or outcome) relative to another group

Type I Error (Alpha Level): The probability of rejecting the null hypothesis when it is true. At the beginning of our hypothesis test, this probability is equal to our level of significance, usually 0.05 or 5%. This means we are willing to take a 5% chance of rejecting the null hypothesis when in fact it is true

Type II Error (Beta Level): The probability of failing to reject the null hypothesis when it is false. In other words, it is the probability of failing to accept a true alternative

Power of the Test: The probability of accepting a true alternative

In this chapter, we will reason with the uncertainty due to sampling variation regarding the value of the four population parameters of interest by asking the following questions:

- What do we mean by the null and alternative hypothesis?
- What do we mean by one-sided and two-sided alternatives?
- How do we reason to statistical decisions regarding the value of population parameters using hypothesis testing?
- What does the **p-value**, the probability value used for making statistical decisions using hypothesis testing, really mean?
- What is meant by the level of significance for the hypothesis test?
- What do we mean by type I and type II errors?
- What do we mean by the power or the test?

We will learn how to reason with the uncertainty regarding the value of our four population parameters of interest by reasoning through the four steps of hypothesis testing. The four steps of hypothesis testing will be very similar in each case, all resulting in a p-value for making a statistical decision. However, we will need to understand how to explain what the p-value means in terms of the particular population parameter of interest.

8.2 NULL AND ALTERNATIVE HYPOTHESES

The **null hypothesis** is a statement regarding the currently accepted (or status quo) value for the population parameter. It could be the hypothesis stating the population mean height of men in the United States is equal to the currently accepted value. It could be the hypothesis the true effect size (or difference between treatment means) is zero in the population. It could be that a particular coin is fair. When it comes to researchers comparing treatments, it is reasonable to begin (before the data is collected) with the assumption that the treatments are equally effective on average.

The aim of the researcher is to challenge the status quo. The researcher believes that their new treatment is more effective than placebo or a competitor's drug. They want to run a test (on the data they will collect) to see if they can find statistical evidence that the true effect size is not equal to zero.

The **alternative (research) hypothesis** is a statement rejecting the null hypothesis. The alternative is our research hypothesis. It could be the hypothesis that the mean height of men in the United States is not equal to the currently accepted value. It could be the hypothesis that the true effect size (or difference between treatment means in the population) is not zero. It could be that a particular coin is not fair. The data will provide the strength of the evidence as to whether or not we should reject the null hypothesis in favor of the alternative hypothesis.

We can think of the null hypothesis as analogous to the presumption of innocence at the beginning of a jury trial. The alternative hypothesis is analogous to the rejection of innocence in favor of a guilty verdict. The researcher is the prosecutor in the trial. No matter what the prosecutor believes, they have to respect the assumption of innocence (the null hypothesis) at the beginning of the trial and work with the evidence to try and convince the jury that the defendant is guilty. Only when the evidence is beyond a reasonable doubt will the jury reject innocence in favor of guilt. If the evidence is not beyond a reasonable doubt, the jury will not say the defendant is innocent. The jury will say the defendant was found to be not guilty.

Our statistical evidence is to be found in the data that we collect. Only when the statistical evidence is beyond a reasonable doubt will we reject the null hypothesis in favor of the alternative. If the statistical evidence is not beyond a reasonable doubt, we will not accept the null hypothesis. We will simply state that we have not found enough statistical evidence in favor of the alternative. Therefore, we fail to reject the null hypothesis.

Due to sampling variation, our sample statistic may deviate somewhat from the null value for the population parameter, and the null hypothesis could still be true. However, at some point, when the sample statistic deviates so far from the null value, we have to question the assumption that the null value is correct and decide whether or not to reject it. We have to decide where to draw the line that represents the border between sticking with the null hypothesis or rejecting it in favor of the alternative. If our sample statistic crosses that borderline, we have moved beyond what is considered a reasonable doubt. We will reject the null hypothesis and state that we have found statistical evidence in favor of our alternative (research) hypothesis.

The most commonly accepted borderline used for making this decision is when our sample statistic deviates from the null value (for the population parameter) by more than two standard errors. When this is the case, the sample statistic is considered too far from the null value to be simply due to sampling variation, given the null value is correct. We say that we have found an accepted level of statistical evidence in our data to reject the null hypothesis in favor of the alternative. Our results are considered statistically significant. There is a lot of discussion (and debate) in the scientific community regarding where the borderline should be drawn for making statistical decisions using hypothesis testing. We will return to this discussion once we have a better understanding of the reasoning of hypothesis testing.

8.3 POPULATION PROPORTION

We will begin to understand the reasoning of hypothesis testing through the example of the coin-toss experiment. In Chapter 5, we talked about how the probability of getting a head is calculated by tossing the coin a very large number of times (an infinite number of times to be exact), and calculating the proportion of heads. From the point of view of statistical analysis of our data, we can think of this probability as our population parameter of interest, the population proportion. It is the true proportion of heads we would expect from tossing a coin a certain number of times. We can think of the actual proportion of heads we obtain as our sample proportion.

Example 8.1

In the (hypothetical) **Example 6.3**, we constructed the sampling distribution of sample proportions of heads from repeated coin-tosses. The sampling distribution was based on samples of 100 coin tosses and assuming that the coin is fair.

We decide to run a hypothesis test to assess whether or not our coin is biased. If the coin is fair, then we expect it to land heads 50% of the time. In other words, the true proportion of heads (the population parameter) we expect to get from repeatedly flipping the coin a very large number of times is 0.5. We set up our hypothesis test as follows:

Null Hypothesis: *population proportion of heads is equal to 0.50 ($H_0: p = 0.50$)*
Alternative Hypothesis: *population proportion of heads is not equal to 0.50 ($H_A: p \neq 0.50$)*

The alternative hypothesis is simply a statement we make if we obtain a sample statistic (a sample proportion of heads) that deviates so far from the null value, it is highly unlikely to occur, given the null value is correct (that the coin is fair).

Due to sampling variation, we know we could get a sample proportion of heads that deviates somewhat from 0.50, even when the coin is fair. From the result of the central limit theorem, we can calculate the range of possible sample proportions we expect, based on the number of times we toss a fair coin.

We decide to toss our coin 100 times and calculate the sample proportion of heads. If the coin is fair, we expect 95% of possible sample proportions (of heads) to be within two standard errors of 0.50. The standard error is equal to 0.05 (see appendix for calculation).

$$\text{null value} \pm 2 \times (\text{standard error})$$
$$0.5 \pm 2 \times (0.05)$$
$$0.5 \pm 0.10$$
$$[0.40, 0.60]$$

If the coin is fair, we expect 95% of possible sample proportions of heads to fall between 0.40 and 0.60, based on 100 coin tosses. If our sample proportion of heads is outside of this range of values, then the deviation of the sample proportion from 0.5 is considered statistically significant. It is unlikely to have been due to sampling variation alone given the coin is fair. We can say that there is statistical evidence in our sample data that the coin is biased toward either heads or tails.

In other words, whenever we toss a coin 100 times, if we get less than 40 heads (0.40) or greater than 60 heads (0.60), we can say that it is highly unlikely that our coin is fair. If the coin is fair, obtaining a total number of heads more extreme than these values would happen only 5% of the time. We would expect only one out of every twenty experiments (or 5%) of this kind to result in a sample statistic that deviates so far from the null value for the population parameter.

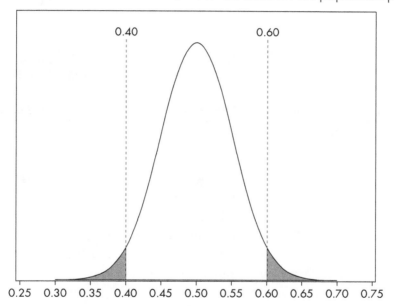

Figure 8.1: Normal Distribution of Possible Sample Proportions (Based on 100 Coin Tosses)

Data source: http://blogs.sas.com/content/sastraining/2014/06/10/producing-normal-density-plots-with-shading/

When our sample statistic is outside this range of values we, can choose between two possible scenarios when making a statistical decision regarding the value of the population parameter. For this example, we can decide that the coin is fair and say that we just experienced an unusual chance event due to sampling variation. Or we can decide that we have found statistical evidence that the coin is biased. It is generally accepted that if the sample statistic deviates so far from the null value that it would occur only 5% of the time or less (assuming that the null value is correct), then the null hypothesis should be rejected in favor of the alternative hypothesis.

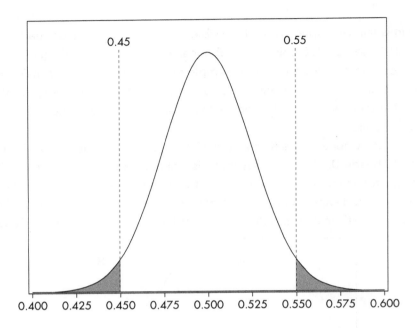

Figure 8.2: Normal Distribution of Possible Sample Proportions (Based on 400 Coin Tosses)

Data source: http://blogs.sas.com/content/sastraining/2014/06/10/producing-normal-density-plots-with-shading/

An analogous real-world example is a very close election race for president where the electorate seem to be split 50:50 between two candidates: candidate A and candidate B. If we decide to run a hypothesis test to challenge this claim, we could start with the null hypothesis that candidate A will get 50% (or 0.50) of the vote. We select a random sample of 100 voters looking for statistical evidence that the percentage of votes candidate A will receive is not equal to 50% (or 0.50), the alternative hypothesis. In our sample of 100 voters, if we find less than 40 voters or greater than 60 voters stating they are going to vote for Candidate A, then this would be considered enough statistical evidence to reject the null hypothesis in favor of the alternative hypothesis. In other words, there is strong indication in the data that Candidate A will not receive 50% of the population vote.

If we increase the number of coin tosses to 400, the standard error decreases by half to be equal to 0.025 (see appendix for calculation). The central limit theorem states that with a larger sample size, we can expect a narrower range of possible sample proportions of heads around the null value of 0.50. If the coin is fair, we expect 95% of possible sample proportions (of heads) to fall between 0.45 and 0.55, based on 400 coin tosses. If our sample proportion falls outside this range of values, then the deviation of the sample proportion from 0.5 is considered statistically significant. It is unlikely to have been due to sampling variation alone, given the coin is fair. We can say we have found statistical evidence that the coin is biased towards either heads or tails.

In other words, whenever we toss a coin 400 times, if we get fewer than 180 heads or greater than 220 heads, we can say that it is very unlikely that our coin is fair. If the coin is fair, obtaining a total number of heads more extreme (or further from 200 heads, the number we expect from a fair coin) than these values would happen only 5% of the time. When this happens, we say we have found enough statistical evidence to reject the null hypothesis in favor of the alternative hypothesis that the coin is biased.

The larger the sample size, the narrower the range of possible sample statistics (around the null value for the population parameter) we expect to obtain. For our coin-toss example, the larger sample size means our sample proportion of heads does not have to deviate as far from the null value for it to be considered statistically significant.

8.4 POPULATION MEAN

In the last section, we discussed the core statistical reasoning used in hypothesis testing for making statistical decisions about the values of population parameters. We will now return to the hypothetical height example from Chapter 7. We will bring more structure to the decision making process by laying out the four steps of hypothesis testing using this example.

Example 8.2

In the (hypothetical) **Example 7.1**, we constructed a confidence interval for the population mean height of men in the United States based on a sample size of 144 men. We will now complete a hypothesis test for the population mean height.

According to the top Google search result (from halls.md) in 2019, the currently accepted population mean height of men in the United States is equal to 70 inches. We wish to challenge this claim using hypothesis testing. We select a random sample of 144 men, resulting in a sample mean of 69.5 inches and a sample standard deviation of 3 inches. The standard error is equal to 0.25 inches (see appendix for calculation).

Our hypothesis test can be performed in four steps. The first step is a statement regarding the value of the population parameter in the null and alternative hypotheses:

Step I – Hypotheses

Null Hypotheses: population mean is equal to 70 inches ($H_0: \mu = 70$)
Alternative Hypothesis: population mean is not equal to 70 inches ($H_A: \mu \neq 70$)

By testing (or challenging) the claim that the population mean is equal to 70 inches, we are stating that we are looking for statistical evidence (using our sample data) in favor of our alternative (research) hypothesis that the population mean is not equal to 70 inches.

Step II – The Model

As long as the necessary assumptions and conditions hold, the central limit theorem states that the distribution of possible sample mean heights follows the normal distribution, centered at the population mean height. For hypothesis testing, we call the normal distribution of sample mean heights the normal model, or simply the model. Under the null hypothesis, we are assuming the model is centered at 70 inches. The question is: what are the chances of getting our sample mean height given the null hypothesis is true?

Using the standard error of 0.25 inches, we can calculate the range of sample mean heights that we would expect, given the population mean height is equal to 70 inches.

If the population mean height is 70 inches, we expect 95% of sample mean heights (based on sample sizes of 144 men) to fall between 69.5 and 70.5 inches. If our sample mean height falls beyond these borderline values, then it is considered statistically significant. It is considered an unusual (or extreme) value, given the null hypothesis is true. Our null hypothesis is rejected in favor of the alternative hypothesis.

Our sample mean height of 69.5 inches (two standard errors below 70 inches) is exactly equal to the borderline value used for making statistical decisions. Our sample mean height is almost enough evidence to reject the null hypothesis in favor of the alternative. However, based on our criteria, it is not far enough away from the null value of 70 inches to reject this claim. We will formulate our decision making by calculating what is known as a p-value (a probability value), and explaining what it means in the context of this example.

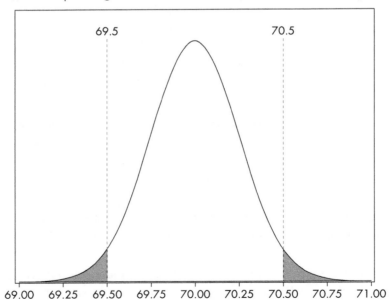

Figure 8.3a: The Model of Possible Sample Mean Heights—Based on Samples of 144 Men

Data source: http://blogs.sas.com/content/sastraining/2014/06/10/producing-normal-density-plots-with-shading/

Step III – Calculations

The value of the p-value is related to the number of standard errors our sample statistic (sample mean height) deviates from the null value for the population parameter (population mean height). The further a sample statistic deviates from the null value (measured in standard errors), the smaller the p-value. The smaller the p-value, the stronger the statistical evidence in favor of the alternative hypothesis.

In Chapter 4, we calculated a standardized score (a z-score), a number that told us how many standard deviations an individual value deviates from the sample mean. The z-score also enabled us to calculate the probability of getting a value more extreme than any particular individual's value.

We are now working with the normal model (or distribution) of possible sample means. Therefore, we can use a similar logic to calculate a standardized score that tells us how many standard errors our sample mean height of 69.5 inches deviates from the null value for the population mean height. As already mentioned, technically, we are working with a t-distribution of sample means and not a normal distribution of sample means. Therefore, the standardized score we are calculating is technically a t-score and not a z-score. We will continue to treat it as a z-score for any calculations we make.

In hypothesis testing, the standardized score is known as the **test statistic**. Once calculated, we can then use the test statistic to calculate our p-value under the standard normal distribution (also discussed in Chapter 4). In this example, the test statistic is calculated as follows:

$$t = \frac{\bar{x} - \mu}{\frac{s}{\sqrt{n}}}$$

test statistic = (sample mean – null value)/standard error
= (69.5 – 70)/0.25
= –2

The gray shaded areas in Figure 8.3b represent the p-value. The p-value is defined as the probability of getting the sample statistic we observed, or one more extreme (even further from the null value), given the null value for the population parameter is correct. In this example, the p-value is the probability of getting a sample mean height of 69.5 inches (or less) or 70.5 inches (or greater), given that the population mean height is equal to 70 inches. A sample mean height of 70.5 inches or greater would be just as much statistical evidence in favor of the alternative hypothesis as a sample mean height of 69.5 inches or less. We need to take both possibilities (or probabilities) into account when making statistical decisions using hypothesis testing. In other words, we are calculating the chances of getting the sample mean height of 69.5 inches (or a sample mean even further from 70 inches), assuming the population mean height is equal to 70 inches.

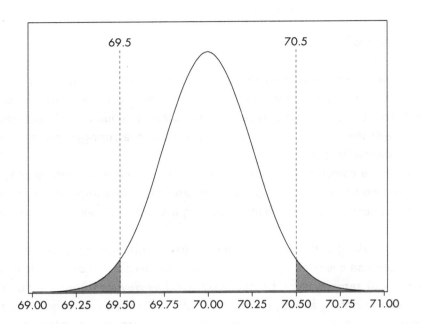

Figure 8.3b: The p-Value for the Hypothesis Test

Data source: http://blogs.sas.com/content/sastraining/2014/06/10/producing-normal-density-plots-with-shading/

Our sample mean height of 69.5 inches (two standard errors below 70 inches) converted to a standardized score (test statistic) equal to −2. The equivalent standardized score for 70.5 inches (two standard errors above 70 inches) is equal to +2. Therefore, our p-value can be calculated as the probability of getting a test statistic less than −2 or greater than +2 under the standard normal distribution. We know from the empirical rule for the normal distribution, this probability is equal to 0.05. In this example, this probability is our p-value for making a statistical decision regarding the value of the population mean height of US men.

Step IV – Conclusion

As previously stated, based on the null hypothesis, a sample statistic will deviate from the null value (for the population parameter) by more than two standard errors only 5% of the time. In other words, when our sample statistic is more than two standard errors away from the null value, the p-value will be less than 0.05.

This borderline value of 0.05 is known as the **level of significance**. When our p-value is less than 0.05, we can say that we have found statistical evidence in our sample data to reject the null hypothesis in favor of the alternative hypothesis.

Our p-value is exactly equal to 0.05 and therefore is not less than the level of significance of 0.05. Therefore, we have almost found (but not quite) enough statistical evidence to reject the null hypothesis that the population mean height is equal to 70 inches (in favor of the alternative), at the 0.05 level of significance. However, since it is not less than 0.05, we fail to reject the null hypothesis.

Some textbooks will state that if the p-value is less than or equal to 0.05, then we should reject the null hypothesis in favor of the alternative. In reality, no p-value will be exactly equal to 0.05 if you write the value out to enough decimal places. It will always be either less than or greater than 0.05. Technically, our exact p-value in this example is less than 0.05 (it is equal to 0.046 to three decimal places), and thus we should have rejected the null hypothesis. However, for ease of explanation, we will leave the p-value at 0.05 and stick with our decision not to reject the null hypothesis.

For the sake of argument, if we were to reject the null hypothesis in favor of the alternative (since our p-value is technically less than 0.05), this does not mean that our sample mean height of 69.5 inches is the value of the population mean height. All we are saying is that we found enough statistical evidence in our sample data to state that the population mean height is not equal to 70 inches. If we started with the null hypothesis (and therefore the model) that the population mean height is equal to 69 inches, we would have come to the same conclusion. The sample mean height of 69.5 inches is just as far from the null value of 69 inches as it is from the null value of 70 inches.

Therefore, under both scenarios (or models) used to analyze the same sample data, we would reject the null hypothesis. In other words, when we reject the null hypothesis, we are rejecting the statistical model that states the population parameter is equal to the null value.

8.5 WHAT DOES THE P-VALUE REALLY MEAN?

P-values are a convenient and much used way for making statistical decisions. Our entire analysis can be boiled down to one value, and if that value is less than 0.05, then we can say we have found statistical evidence in favor of our alternative hypothesis. Our results are considered to be statistically significant.

How the p-value is used for making statistical decisions can give the illusion of certainty to a process that is inherently uncertain. If the sample size is small, then the variation in possible sample statistics around the unknown population parameter (measured using the standard error) can be quite large, meaning that there is still a lot of uncertainty regarding the value of the population parameter. A p-value does not reflect this uncertainty, whereas a confidence interval does. The p-value simply tells us the probability of obtaining the sample statistic we observed (or one more extreme), given the null hypothesis is true.

The level of significance of 0.05 is simply a borderline value, a probability value used for making a statistical decision using hypothesis testing. It means that at the beginning of our analysis we are willing to take a 5% chance of rejecting the null hypothesis when it is actually true.

This is because 5% of the time we expect to obtain a sample statistic that deviates from the null value by more than two standard errors when the null hypothesis is true. For example, let's assume the population mean height of men in the United States is in fact 70 inches. We select numerous samples of 144 men and calculate the sample mean heights. If we repeat the process enough times, we will find approximately 5% of the sample mean heights to be more than two standard errors (or 0.5 inches) from 70 inches. Due to sampling variation, there is a 5% probability we will obtain a sample mean height that rejects the null hypothesis in favor of the alternative, when in fact the null hypothesis is true.

It is important to understand that there is more than one place where the borderline can be drawn. Research includes the analysis of data from different fields with different consequences for rejecting the null hypothesis when it is actually true. Where the borderline is drawn for making statistical decisions should be decided on a case-by-case basis.

For example, if a drug has shown (in earlier experiments) to have serious side effects on some patients, the researcher should use a stricter level of significance, for example 0.01 before stating that the drug is effective. This means that the researcher is only willing to take a 1% chance of stating that the drug is more effective (than say a placebo or comparative drug), when in fact it is no more effective. The p-value needs to be less than 0.01 for the researcher to state they have found statistical evidence that the drug is effective. If the consequences of bringing the drug to market means that a small but significant percentage of individuals will suffer serious side effects, we need to be very confident that the effect size found in the sample data points to a real effect of the drug in the population and not simply due to sampling variation.

However, the opposite should also be true when it comes to considering whether differences in the rate of adverse events across treatments is statistically significant. For example, at the end of our final chapter titled Integrity in Research, we will discuss the drug Vioxx, taken off the market in 2004 because it was found to be causing heart attacks. In a large study with around 8,000 patients, Vioxx was found to causing four times as many heart attacks compared to a drug called Aleve. Even though the p-value was not presented in the journal article, it was less than 0.05. However, even if the p-value was greater than 0.05 (say 0.10 or 0.15), any difference in the rate of heart attacks across treatments should have been cause for concern. The chances that we make the wrong decision (conclude Vioxx causes heart attacks when it does not) should not need to be so small (less than 0.05) when it comes to adverse events like heart attacks. If there was any chance the drug was causing heart attacks, that should have been cause for concern. The line in the sand for statistical significance should be balanced against the consequences of making the wrong decision. This was a well conducted study with a large sample size resulting in a small standard error. The chances are the sample statistics calculated from this data were not very far from the truth. The drug was on the market for four years resulting in a population of 80 million people taking the drug. As it turned out, the rate of heart attacks on Vioxx found in this study was very close to the truth.

Sir Ronald Fisher stated clearly in his 1956 book titled *Statistical Methods and Scientific Inference* that a statistical decision should be decided on a case-by-case basis:

> *A man who 'rejects' a hypothesis provisionally, as a matter of habitual practice, when the significance is 1% or higher, will certainly be mistaken in not more than 1% of such decisions ... However, the calculation is absurdly academic, for in fact no scientific worker has a fixed level of significance at which from year to year, and in all circumstances, he rejects hypotheses; he rather gives his mind to each particular case in the light of his evidence and his ideas.*

The concept of using a single level of significance for all hypothesis tests is no less absurd today than it was when Ronald Fisher first wrote these words. However, it is standard practice when a p-value is less than 0.05 to reject the null hypothesis in favor of our alternative (research) hypothesis and declare statistical significance. In any hypothesis test, we are choosing between two possibilities. Therefore, there are two possible errors we could make in our decision making process, which we will discuss in-depth at the end of this chapter. The probability that we made an error in our decision making should be balanced against the cost and benefits of the treatment, and the possible consequences for making the wrong decision. The p-value should not be simply compared to the same line in the sand every time a hypothesis test is conducted.

The good news is that there has been some movement in the statistical community towards the idea of eliminating the use of the term statistical significance altogether. In 2019, a commentary titled "Scientists rise up against statistical significance" published in the journal Nature, well-respected statisticians and researchers discuss the problems caused in research by having a set threshold for statistical significance. They discuss some of the problems with significance testing and suggest a greater emphasis on understanding and reasoning with confidence intervals and lesser emphasis on the p-value.

Again, in 2019, the journal The American Statistician in a journal article titled "Abandon Statistical Significance", well-respected statisticians and researchers make a well-reasoned argument for abandoning statistical significance and putting the p-value in its proper place in the decision making process. They discuss the many errors and biases that can be an inherent part of the p-value calculation including:

> *... any and all forms of systematic or nonsampling error which vary by field but include measurement error; problems with reliability and validity; biased samples; nonrandom treatment assignment; missingness; nonresponse; failure of double-blinding; noncompliance; and confounding.*

COMMENTARY: SCIENTISTS RISE UP AGAINST STATISTICAL SIGNIFICANCE

Web Link: https://www.nature.com/articles/d41586-019-00857-9

Search Term: Scientists rise up against statistical significance

JOURNAL ARTICLE: ABANDON STATISTICAL SIGNIFICANCE

Web Link: https://www.tandfonline.com/doi/full/10.1080/00031305.2018.1527253

Search Term: Abandon Statistical Significance

If the data is of poor quality and does not adhere to the necessary assumptions and conditions for analysis, the end result, the p-value, will be of poor quality. In their summary, they recommend a more holistic view of the evidence:

> Further, each is a purely statistical measure that fails to take a more holistic view of the evidence that includes the consideration of the currently subordinate factors, that is, related prior evidence, plausibility of mechanism, study design and data quality, real world costs and benefits, novelty of finding, and other factors that vary by research domain.

What the commentary and journal article both agree on is that it is time to abandon the use of the threshold or line in the sand used for declaring statistical significance. They are not suggesting that the use of p-values should be eliminated altogether. However, they all agree we have to move away from the illusion of certainty created by the lines in the sand used for declaring statistical significance. We need to let the waves of uncertainty (that are our statistical models) wash away the lines in the sand for good. In other words, when it comes to statistical measures we need to focus more on the sample statistic: our estimate of truth; the standard error: how far we expect our sample statistic to be from the truth; the confidence interval and the probability that the truth is within its limits. In other words, we need to face, embrace and reason with the uncertainty due to sampling variation that is an inherent part of the process, and not simply fall into the illusion (and comfort) of certainty that declaring (or not declaring) statistical significance provides.

Truth is a pursuit, not a destination. When we decide to pursue truth using data and statistics we have to understand that at the end of the process we have proven nothing. The process may have brought us closer to the truth but not to truth itself. In other words, the answer is not black or white. The answer lies in the gray area of uncertainty.

When it comes to explaining what the p-value really means, even seasoned researchers find it difficult to explain in plain English. The following article from the website FiveThirtyEight includes a short video where a number of researchers are asked to explain what the p-value means:

> Not Even Scientists Can Easily Explain P-Values
>
> Last week, I attended the inaugural METRICS conference at Stanford, which brought together some of the world's leading experts on meta-science, or the study of studies. I figured that if anyone could explain p-values in plain English, these folks could. I was wrong.

To be clear, everyone I spoke with at METRICS could tell me the technical definition of a p-value—the probability of getting results at least as extreme as the ones you observed, given that the null hypothesis is correct—but almost no one could translate that into something easy to understand.

When thinking about what the p-value really means as it relates to the null value for the population parameter, the meaning of the p-value is quite clear. It is the probability of obtaining our sample statistic (or one more extreme), given the null value for the population parameter is correct.

It is when we have to reason with the p-value to choose between the null and alternative hypothesis (by comparing it to the level of significance), that the mental challenge can feel quite exhausting. The p-value is a convenient way of making statistical decisions but thinking about what it means in this context can be challenging (and confusing).

When reasoning through the (hypothetical) height example in the last section, we stated the further a sample mean deviates from the population mean, the smaller the p-value. The smaller the p-value, the stronger the evidence in favor of the alternative hypothesis. However, what does this actually mean?

If we obtain a sample mean much greater than the null value for the population mean (resulting in a very small p-value), it is highly likely that the actual value for the population mean is greater than the null value. If we obtain a sample mean much smaller than the null value for the population mean (resulting in a very small p-value), it is highly likely that the actual value for the population mean is less than the null value. How much greater than or less than the null value we can't say for certain. However, what we can say is that the smaller the p-value, the further the population mean is likely to be from the null value. Including a confidence interval along with our p-value provides a range of plausible values for the population parameter.

The key takeaway from this discussion is that we should not try too hard to grasp the meaning of the p-value. It is a convenient calculation for making statistical decisions with hypothesis testing but not an ideal one. We should simply think about the meaning of the p-value more loosely as an indication of the strength of the evidence in favor of the alternative hypothesis.

An analogy that could help you master the four steps of hypothesis testing is to compare the four step process to the challenge of playing the four chords to your favorite song on a guitar for the first time. The song might have four chords—like C, F, G and Am—analogous to the four steps of hypothesis testing.

When you first sit down to play the song on a guitar, you may feel that this will not take very long to master. It is just four simple chords you need to move your fingers between and play.

NOT EVEN SCIENTISTS CAN EASILY EXPLAIN P-VALUES

Web Link: http://fivethirtyeight.com/features/not-even-scientists-can-easily-explain-p-values/

Search Term: Not Even Scientists Can Easily Explain P-Values

THE ASA'S STATEMENT ON P-VALUES: CONTEXT, PROCESS, AND PURPOSE

Web Link: http://amstat.tandfonline.com/doi/abs/10.1080/00031305.2016.1154108

Search Term: ASA's Statement on P-Values

However, you will soon experience that even getting your fingers to play the first chord is very challenging. It takes real focus and effort to just get that one chord sounding right.

Hypothesis testing challenges your mind in the same way learning a song on a guitar challenges your fingers. However, with enough patience and repetition, your mind will gain a firm grip of each of the four steps in the process and learn how to reason (or move smoothly) between them.

We need to reason through the steps of hypothesis testing slowly and thoughtfully. The p-value is the end result of a very challenging process of reasoning with uncertainty due to sampling variation in order to make a statistical decision regarding the value of the population parameter. Due to the continuing controversy over the misuse and misunderstanding of p-values, the American Statistical Association *in 2016* published a statement regarding their use titled *The ASA's Statement on p-Values: Context, Process, and Purpose*.

The statement discusses what the p-value measures (and what it does not measure) and how it can get misused or misunderstood. As in the journal article titled "Abandon Statistical Significance", the statement points out that:

> *Practices that reduce data analysis or scientific inference to mechanical "bright-line" rules (such as "p < 0.05") for justifying scientific claims or conclusions can lead to erroneous beliefs and poor decision making. A conclusion does not immediately become "true" on one side of the divide and "false" on the other. Researchers should bring many contextual factors into play to derive scientific inferences, including the design of a study, the quality of the measurements, the external evidence for the phenomenon under study, and the validity of assumptions that underlie the data analysis.*

Statistics, the discipline, is a rigorous and challenging craft. It takes a lot of hard work to minimize the distance between our estimate of truth, the sample statistic, and the truth itself. It is very encouraging that influential members of the statistical community are moving towards emphasizing a more holistic view of the evidence, and encouraging researchers to reason to a greater degree with the uncertainty in the process of statistical or scientific inference.

In our final chapter titled Integrity in Research, we will see how p-values can get misused or misunderstood to the researcher's advantage. One way this can occur is by choosing a one-sided alternative over a two-sided alternative.

8.6 ONE-SIDED VERSUS TWO-SIDED ALTERNATIVES

In our (hypothetical) height example, we conducted what is known as a **two-sided alternative** hypothesis test. This means that we are open to our data pointing to statistical evidence that the population mean height of men in the United States is less than 70, or that the population mean height of men in the United States is greater than 70 inches. In other words, we are open to the possibility that the statistical evidence could point in either direction.

Say, from your observations (around your college or neighborhood), you think the population mean height of men is less than 70 inches. As a result, you might be tempted to run what is known as a **one-sided alternative** hypothesis test: simply looking for statistical evidence that the population mean height is less than 70 inches.

Using this alternative, we can analyze the same random sample of 144 men's heights, (see Example 6.1) with a sample mean of 69.5 inches, a sample standard deviation of 3 inches, and a standard error of 0.25 inches.

Step I – Hypotheses

Null Hypothesis: population mean is equal to 70 inches ($H_0: \mu = 70$)
Alternative Hypothesis: population mean is less than 70 inches ($H_A: \mu < 70$)

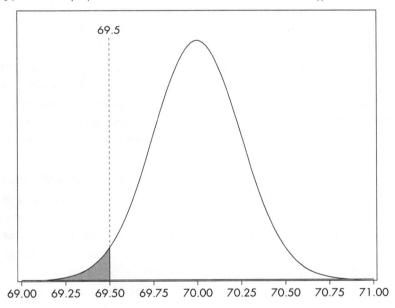

Figure 8.4: The P-Value for the Hypothesis Test

Data source: http://blogs.sas.com/content/sastraining/2014/06/10/producing-normal-density-plots-with-shading/

All the remaining steps in this analysis remain the same except for how the p-value is calculated. The grey shaded area in Figure 8.4 represents the p-value. Since we are only looking for statistical evidence (against the null hypothesis) in one direction, our p-value is calculated as the probability of getting a sample mean of 69.5 inches or less. The p-value is cut in half, equal to 0.025, and therefore statistically significant at the 0.05 significance level.

The benefit of choosing a one-sided alternative is that it gives our test greater power (a concept we will discuss at the end of this chapter) to find statistical evidence in favor of what we are looking for. In this case, we found statistical evidence in favor of our alternative that the population mean height is less than 70 inches; whereas in our original hypothesis test, we did not.

However, there is a danger in choosing a one-sided alternative. By gaining greater power to find what we are looking for, we lose the ability to detect statistical evidence pointing in the other direction. For example, if our sample mean height turned out to be equal to 70.5 inches, based on our one-sided alternative (population mean is less than 70 inches), the p-value is calculated as the probability of getting a sample mean height of 70.5 inches or less. In this case, the p-value would be equal to 0.975, the area under the normal curve (in Figure 8.4) to the left of 70.5 inches. The sample mean height of 70.5 inches is just as much statistical evidence against the null hypothesis as the original sample mean of 69.5 inches (since both values deviate from 70 inches by a half an inch or two standard errors). However, our one-sided alternative hypothesis test will fail to detect this fact.

In many applications of statistics (data science, engineering, agriculture, finance, climate change), the researcher may only be interested in looking for statistical evidence in one direction. For example, a company might only be interested in finding evidence that the mean lifetime of their lightbulbs is greater than 700 hours. They will likely choose the one-sided alternative that the population mean is greater than 700 hours. The researcher simply needs to be aware that their test will have greater power (to detect what they are looking for) by choosing a one-sided alternative; and that they will lose the ability to detect whether the mean lifetime of the bulbs is less than 700 hours.

When it comes to testing the effectiveness of a drug, say, for lowering blood pressure, it can be dangerous to choose a one-sided alternative. In early studies, based on a small sample of patients, the sample data may show statistical evidence that the drug is effective at lowering blood pressure. In follow-up studies, based on a larger sample size, the researchers might be tempted to conduct a one-sided alternative hypothesis test; simply looking for further statistical evidence that this drug is effective at lowering blood pressure.

However, as we have discussed, small samples can result in highly variable sample effect sizes, due to large sampling variation. The smaller sample size may show statistical evidence the drug is effective at lowering blood pressure; whereas the larger sample may show statistical evidence that it increases blood pressure. In the larger study, if the researcher runs a one-sided alternative hypothesis test—only looking for evidence that the drug is effective at lowering blood pressure—their test will fail to detect evidence pointing to the conclusion that the drug actually *increases* blood pressure.

When reading the results of a study, it is important to be aware of this issue. A hypothesis test based on a one-sided alternative might result in a statistically significant p-value equal to 0.03, for example. However, the p-value based on a two-sided alternative would equal 0.06, a non-statistically significant result. If the researcher used a one-sided alternative and you don't agree with their reasoning for doing so, make the necessary adjustments to the p-value and draw your own conclusions.

After you run your hypothesis test, you might be tempted, based on your sample data, to change your one-sided alternative hypothesis to point in the other direction. However, you should always choose your null and alternative before you collect your data. It is poor statistical practice to do otherwise.

You should always question the use of a one-sided alternative in research studies. The researcher may justify its use by stating that they were only interested in whether or not their drug was more effective that their competitor's drug or a placebo treatment. However, what about the possibility that it is less effective? When pursuing truth with data and statistics, we should look for what is true and not what we want to be true! In our final chapter on Integrity in Research, we will look at a controversial study in psychology where the researcher used a one-sided alternative to ensure his claims were considered statistically significant.

8.7 POPULATION MEAN DIFFERENCE

We will now apply the reasoning of hypothesis testing to situations where we are comparing groups of individuals with regard to a particular quantitative variable. We will complete what is known as a comparison of means hypothesis test. The four-step reasoning of hypothesis testing will be exactly the same as previously shown for determining the value of the population mean. Only the calculation of the test statistic to obtain the p-value will be different because we are analyzing two groups of (quantitative) data instead of one.

Example 8.3

Let's return to the NHANES study weight data to complete a comparison of means hypothesis test. This test will complement the confidence interval for the population mean difference in weight between men and women we calculated in **Example 7.5**. Our null and alternative hypotheses are as follows:

Step I – Hypotheses

Null Hypothesis: population mean difference is equal to zero kg ($H_0: \mu_M - \mu_F = 0$)
Alternative Hypothesis: population mean difference is not equal to zero kg ($H_A: \mu_M - \mu_F \neq 0$)

The null hypothesis is simply a statement of equality of population means. From the confidence interval we calculated in the last chapter, we have little to no doubt that the population mean weight of men is different from the population mean weight of women. However, in studies where the researchers are making treatment comparisons, the null hypothesis should be that the population means are the same (the population mean difference is equal to zero). At stated previously, this is analogous to the assumption of innocence in a jury trial. The prosecutor must respect this assumption at the beginning of the trial no matter what they believe regarding the defendant. It is a reasonable assumption to begin the analysis of your data from the point of view that the treatments are equally effective on average. The data provides the statistical evidence as to whether or not the null hypothesis should be rejected in favor of the alternative that the population means are not the same (or that the treatments are not equally effective on average).

Step II – The Model

As long as the necessary assumptions and conditions hold, the central limit theorem states that the distribution of sample mean differences (men minus women) follows the normal model, centered at the population mean difference. Under the null hypothesis, we are assuming the model is centered at 0 kg. We are starting with the assumption that there is no difference between the mean weight of men and women in the population.

The following table gives a breakdown of the sample data and sample statistics for both groups.

Table 8.1

Sample Groups	Male	Female
Sample Size	2,742	2,794
Sample Mean	85	75
Sample Standard Deviation	20.6	20.9

From Table 8.1, we can calculate the standard error of the sample mean difference, which is equal to 0.56 kg (see appendix for calculation). Using the standard error, we can calculate the range of sample mean differences we would expect, given the population mean difference is equal to 0 kg.

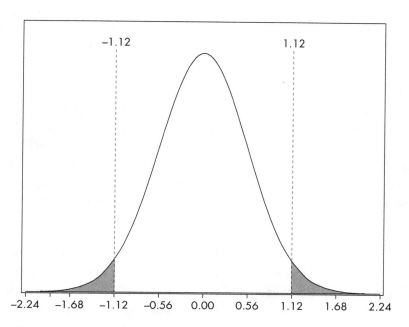

Figure 8.5: The Model of Possible Sample Mean Differences in Weight (Based on Total Sample Size of 5,536 Men and Women)

Data source: http://blogs.sas.com/content/sastraining/2014/06/10/producing-normal-density-plots-with-shading/

If the population mean difference in weight is 0 kg, we expect 95% of sample mean differences in weight to fall between −1.12 kg and 1.12 kg. With such a large sample size, we expect a very narrow range of sample mean differences in weight around 0 kg, given the null hypothesis is correct. The sample mean difference in weight between men and women of 10 kg lies well beyond these borderline values.

Step III – Calculations

$$t = \frac{(\bar{x}_M - \bar{x}_F) - (\mu_M - \mu_F)}{se(\bar{x}_M - \bar{x}_F)}$$

test statistic = (sample mean difference − null value)/standard error of difference
= ((85 − 75) − 0)/0.56
= 10/0.56
= 17.9

The test statistic of 17.9 tells us that the sample mean difference in weight of 10 kg is 17.9 standard errors greater than the null value for the population mean difference of 0 kg.

The p-value is so small that we can't visualize it under the model of possible sample mean differences. The p-value is practically zero, well below the level of significance of 0.05.

Step IV – Conclusion

We reject the null hypothesis in favor of the alternative hypothesis, which states that the population mean difference in weight between men and women is not equal to 0 kg. The very small p-value is very strong statistical evidence that the population mean difference (in weight between men and women) does not equal 0 kg. However, it is the confidence interval that gives us a range of plausible values for that difference. As we saw in **Example 7.5**, we are 95% confident that the population mean difference in weight lies somewhere between 8.88 kg and 11.12 kg, well above 0 kg.

In a statistical report, a confidence interval should always be presented with the p-value so that the reader can have a better sense of the range of plausible values for the population mean difference. There is a duality between a two-sided alternative hypothesis test using a 0.05 (or 5%) level of significance, and a 95% confidence interval. For this type of hypothesis test, if the p-value is less than 0.05, then the 95% confidence interval will not include zero. If the p-value is greater than 0.05, then the 95% confidence interval will include zero. In other words, we can know by looking at the range of plausible values for the population mean difference presented within the 95% confidence interval, whether or not we would reject the null hypothesis in favor of the two-sided alternative at the 0.05 level of significance.

In the *Fast Diet* study we looked at in Chapter 2, the sample mean weight loss was 6.4 kg on the IER diet, and 5.6 kg on the CER diet, a sample mean difference in weight loss of 0.8 kg. With the p-value equal to 0.40, the sample mean difference in weight loss was not found to be statistically significant.

What this means is that (based of the sample size of 107 women), there is a 40% probability of getting a sample mean difference in weight loss of 0.8 kg or greater (or –0.8 kg or less), given the assumption both diets are equally effective (on average) for the entire population. The IER dieters lost more weight (on average) than the dieters on the CER diet. However, this difference in weight loss is not enough statistical evidence to suggest that the IER diet is a more effective weight loss program. As was the case when discussing the confidence intervals presented in this study, it is questionable whether the data adhered to the necessary assumptions and conditions for analysis.

In Chapter 7, we discussed the results of the study titled "Relationship of Collegiate Football Experience and Concussion with Hippocampal Volume and Cognitive Outcomes," first discussed in Chapter 2. We discussed the confidence intervals presented for the sample mean differences in hippocampal volume between the groups the researchers were comparing. The researchers also presented p-values. All three p-values were based on a null hypothesis of no difference in the population mean hippocampal volume for the groups being compared. In other words, the researchers were testing to see if there is statistical evidence that any of the population effect sizes differ from zero across groups. The researchers also presented a statistic known as **Cohen's d** calculated as follows:

$$d = \frac{\bar{x}_1 - \bar{x}_2}{(s_1 + s_2)/2}$$

$$\text{Cohen's d} = \frac{\text{Mean of Group 1 minus Mean of Group 2}}{\text{Pooled Standard Deviation}}$$

Cohen's d is a standardized score that measures the magnitude of the sample effect sizes across groups. It is a useful calculation to include along with the p-value to give a better sense of the magnitude of the effect sizes observed. As a rule of thumb, a Cohen's d equal to 0.2 is considered a small effect size, a value of 0.5 is considered moderate, and a value of 0.8 or greater is considered a large effect size.

Controls versus Players with Concussion

The sample mean difference (controls minus players with concussion) in hippocampal volume was 1,788 uL with Cohen's d equal to 2.15 and a p-value less than 0.001. This means that there is a very small probability of obtaining a sample mean difference in hippocampal volume of 1,788 uL, given that population mean difference in hippocampal volume is zero uL. The lower limit of the 95% confidence interval for the difference [1,317, 2,258] is well above zero uL.

In other words, there is strong statistical evidence of a difference in the mean hippocampal volume (in the population) between controls and players with concussion; and we are highly confident that difference is a value between 1317 uL and 2258 uL.

Controls versus Players without Concussion

The sample mean difference (controls minus players without concussion) in hippocampal volume was 1,027 uL with Cohen's d equal to 1.23 and a p-value less than 0.001. This means that there is a very small probability of obtaining a sample mean difference in hippocampal volume of 1,027 uL, given that population mean difference in hippocampal volume is zero uL. The lower limit of the 95% confidence interval for the difference [556, 1,498] is well above zero uL.

In other words, there is strong statistical evidence of a difference in the mean hippocampal volume (in the population) between controls and players without concussion; and we are highly confident that difference is a value between 556 uL and 1,498 uL.

Players without Concussion versus Players with Concussion

The sample mean difference (players without concussion minus players with concussion) in hippocampal volume was 761 uL with Cohen's d equal to 0.90 and a p-value equal to 0.003. This means that there is a very small probability of obtaining a sample mean difference in hippocampal volume of 761 uL, given that population mean difference in hippocampal volume is zero uL. The lower limit of the 95% confidence interval for the difference [280, 1,242] is well above zero uL.

In other words, there is strong statistical evidence of a difference in the mean hippocampal volume (in the population) between players without concussion and players with concussion; and we are highly confident that difference is a value between 280 uL and 1,242 uL.

Both the hypothesis tests and confidence intervals show strong statistical evidence of a relationship between football playing and hippocampal volume. As was the case when discussing the confidence intervals, the results of the analysis should be cause for concern and further research.

8.8 ANALYSIS OF VARIANCE (ANOVA)— COMPARING SEVERAL MEANS

An alternative but equivalent approach for comparing means is known as analysis of variance or ANOVA as it is often written. ANOVA is a statistical technique most often used for comparing means across more than two groups. For example, in a study comparing three treatments, an ANOVA test will have the following null and alternative hypothesis:

Null Hypothesis: population means are all equal (H_0: $\mu_1 = \mu_2 = \mu_3 = \mu$)
Alternative Hypothesis: at least two of the population means, μ_1, μ_2, μ_3 differ from each other

ANOVA is also known as a global test. When comparing more than two means, ANOVA is testing to see whether there is statistical evidence that any of the population means differ from each other. If the null hypothesis is rejected in favor of the alternative, we can say we have found statistical evidence that at least two of the population means differ from each other.

When ANOVA is used for simply comparing two means, the resulting p-value will be the same as what we would obtain by conducting the comparison of means test we learned about in our last section. In Chapter 2, we critiqued a study that looked at the effect of Botox on a person's ability to pick up on the emotions of others. The researcher compared patients receiving Botox injections to patients receiving a dermal filler called Restylane. The researchers compared the mean percentage of times the patients chose the correct emotion in what is known as the

Reading the Mind in the Eyes Test (RMET). The mean percentage was 69.91 in the Botox group and 76.92 in the Restylane group. When comparing the mean percentages for both groups, the researchers conducted an ANOVA test that resulted in a test statistic known as the F-statistic.

An ANOVA test (for comparing two or more groups) is used to see whether the variability (or difference) in the sample means the mean of the sample means (known as between group variability) is large relative to the variability of the individual measurements around the sample means for each group (known as within group variability). If the measure of between group variability (known as the mean square for treatment group) is large relative to our measure of the within group variability (known as the mean square for error), then the statistical evidence is pointing to the possibility that at least two of the population means differ from each other. The F-statistic is simply the mean square for treatment over the mean square for error. In other words, if the sample means are far away from each other together with a tight cluster of individual measurements around each sample mean, this will result in a large F-statistic pointing to statistical evidence that the population means differ from each other. Note that the mean square is simply another name for variance, the statistic we calculated in Chapter 4 as a stepping stone to calculating the standard deviation.

Since the F-statistic is the result of putting one measure of variability over another, it will have a positive value. The larger the F statistic, the smaller the p-value, and, therefore, the stronger the evidence in favor of the alternative hypothesis that at least two of the population means differ from each other. The F-statistic is simply telling us that the between group variability is large relative to the within group variability. In this Botox example, the F-statistic was equal to 4.33, resulting in a p-value equal to 0.046. The difference of 7% (sample effect size) between these sample statistics was found to be statistically significantly different from a zero effect size. There is statistical evidence at the 5% significance level that there is a difference in the population mean percentage of correct answers on the RMET test when comparing Botox users with Restylane users.

In our final chapter titled Integrity in Research, we will critique the results of research into what the researcher termed power posing. The researcher compared "high-power posers" to "low-power posers" on their willingness to take risks but also compared both groups on testosterone and cortisol levels. We will see that the researcher conducted an ANOVA test to compare means levels of testosterone and cortisol for both groups although a simple comparison of (two) means test would have been sufficient and resulted in the same p-value.

When comparing three or more groups using an ANOVA test, if the test result turns out to be statistically significant, then the next step is to conduct multiple comparisons of (two) mean tests to determine where the real differences in population means lie across groups. However, it is very important that we adjust (increase) the p-values resulting from these tests for the fact that we are making multiple comparisons. As we have learned, every time we conduct a comparison of means test, we are willing to take a 5% chance of rejecting the null hypothesis when, in fact, it is true. In other words, we are willing to take a 5% chance of saying the population means are different when, in fact, they are not. When the ANOVA test rejects the null hypothesis in favor of the alternative, the more treatment groups we have, the more comparison of (two) means tests we will have to run. The more tests we run, the more likely we are going to find a statistically significant result simply by chance alone. In other words, besides finding the two treatment groups that have different population means, we are likely to find statistical evidence that two population means differ when in reality they don't. This can make it impossible to know where the true differences lie. To address this issue, p-values should be adjusted (upwards) for the fact that we are making multiple comparisons of two groups.

There are different methods for adjusting p-values for multiple comparisons, some more conservative than others. Two of the most well-known are the Bonferroni method and Tukey's method. A more in-depth discussion of these methods is beyond the scope of this book. In our final chapter, we will also critique the results of a controversial study on ESP (extrasensory perception) reported in the media. We will discuss the fact that the researcher conducted numerous hypothesis tests without making any adjustments to the resulting p-values. This example is a little different from what we have been discussing here. The researcher did not run an ANOVA test. In fact, the researcher ran a large number of multiple comparisons until he found the answer he was looking for. However, he made no adjustments to the p-values to allow for the possibility of finding a statistically significant result by chance alone. Through this example, we will gain a better understanding of why it is important to adjust our p-values when making multiple comparisons.

In the *New York Times* article titled "Regular Exercise May Keep Your Body 30 Years 'Younger,'" the reporter discusses the results of research that found that regular exercises throughout a person's life can keep the body in much better shape than what you might expect from not doing so. In the related journal article titled "Cardiovascular and skeletal muscle health with lifelong exercise," presented in the journal Applied Physiology, the researcher compared three groups using ANOVA: Lifelong Exercisers (LLE), Older Healthy Nonexercisers (OH), and Young Exercisers (YE). The researchers used Tukey's method to adjust for multiple comparisons of two groups. The following is a summary of their findings:

> A hierarchical pattern was observed (YE > LLE > OH) in $\dot{V}O_{2max}$, ventilation, and heart rate. Absolute $\dot{V}O_{2max}$ (l/min) in YE was 68% greater than LLE and 131% greater than OH ($P < 0.05$) whereas LLE was 37% greater than OH ($P < 0.05$). Similarly, $\dot{V}O_{2max}$ relative to body mass (ml·kg^{-1}·min^{-1}) in YE was 69% greater than LLE and 144% greater than OH ($P < 0.05$) whereas LLE was 44% greater than OH ($P < 0.05$). The YE cohort had a maximal heart rate of 24 beats/min (15%) greater than LLE and 36 beats/min (24%) greater than OH ($P < 0.05$) whereas LLE was 12 beats/min (8%) greater than OH ($P < 0.05$).

NEWS ARTICLE: REGULAR EXERCISE MAY KEEP YOUR BODY 30 YEARS 'YOUNGER'

Web Link: https://www.nytimes.com/2018/11/21/well/move/regular-exercise-may-keep-your-body-30-years-younger.html

Search Term: Regular Exercise May Keep Your Body 30 Years 'Younger'

RELATED JOURNAL ARTICLE: CARDIOVASCULAR AND SKELETAL MUSCLE HEALTH WITH LIFELONG EXERCISE

Web Link: https://www.ncbi.nlm.nih.gov/pubmed/30161005

Search Term: Cardiovascular and skeletal muscle health

The actual results may not be as inspiring as the headline of the news article suggests. However, when comparing older healthy nonexercisers to lifelong exercisers, the study results show statistical evidence of real benefits from a lifetime dedication to exercise. A critique of this study is beyond the scope of this discussion. However, it should be pointed out that the sample sizes were small, and the data were observational in nature. There are likely to be other confounding factors that contribute to the differences in measurements between the three groups besides their exercise habits.

A more in-depth discussion of ANOVA is beyond the scope of this book. However, we have gained enough of a conceptual understanding to complete our own ANOVA test. In our next chapter on linear regression, we will discuss why the results of a multiple linear regression analysis include an ANOVA table. We will learn how to read an ANOVA table and how the table is related to the results of a multiple regression analysis.

8.9 POPULATION PROPORTION DIFFERENCE

We will now apply the reasoning of hypothesis testing to situations where we are comparing groups of individuals with regard to a particular categorical variable. We will complete what is known as a comparison of proportions hypothesis test. The four-step reasoning of hypothesis testing will be exactly the same. Only the calculation of the test statistic to obtain the p-value will be different.

Example 8.4

In the (hypothetical) **Example 7.7**, we constructed a confidence interval for the population proportion difference between men and women who will vote for a particular candidate. The confidence interval was based on a total sample size of 800 men and women. We will now complete a hypothesis test for the population proportion difference.

In this analysis, we are looking for statistical evidence as to whether there is a difference in the proportion of men and women who will vote for a particular candidate in an election. Our null and alternative hypothesis are as follows:

Step I – Hypotheses

Null Hypothesis: population proportion difference equal to zero (**H_0**: $p_M - p_F = 0$)
Alternative Hypothesis: population proportion difference not equal to zero (**H_A**: $p_M - p_F \neq 0$)

Step II – The Model

As long as the necessary assumptions and conditions hold, the central limit theorem states that the distribution of sample proportion differences follows the normal model, centered at the population proportion difference. Under the null hypothesis, we are assuming the model is centered at zero. We are starting with the assumption that there is no difference in the proportion of men and women in the population who will vote for the particular candidate.

We selected a sample of 400 men and 400 women. We found that 220 of the men said they would vote for the particular candidate, a sample proportion equal to 0.55. We found that 180 of women said they would vote for the particular candidate, a sample proportion equal to 0.45. The sample proportion difference is 0.10 (0.55 minus 0.45).

When comparing proportions with hypothesis testing, there is a slight modification to how we calculate the standard error (as compared to how it was calculated when constructing the confidence interval in the last chapter). Please see the appendix for a discussion of this modification and the calculation of the standard error. For this example, it turns out that even with this modification, the standard error is still equal to 0.035 (to three decimal places).

Using the standard error, we can calculate the range of sample proportion differences we would expect, given the population proportion difference is equal to zero.

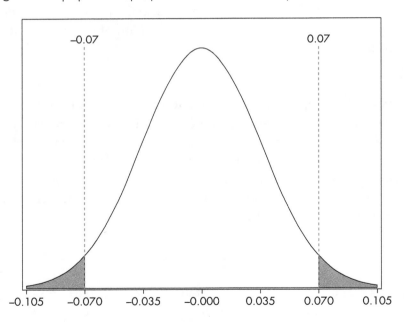

Figure 8.6a: The Model of Possible Sample Proportion Differences

Data source: http://blogs.sas.com/content/sastraining/2014/06/10/producing-normal-density-plots-with-shading/

If the population proportion difference is zero, we expect 95% of sample proportion differences to fall between –0.07 and 0.07. The sample proportion difference (men minus women) of 0.10 lies beyond these limits.

Step III – Calculations

$$z = \frac{(\hat{p}_M - \hat{p}_F) - (p_M - p_F)}{se(\hat{p}_M - \hat{p}_F)}$$

test statistic = (sample proportion difference – null value)/standard error of the difference
= ((0.55 – 0.45) – 0)/0.035
= 0.10/0.035
= 2.86

The test statistic of 2.86 tells us that the sample proportion difference of 0.10 is 2.86 standard errors greater than the null value of zero. The gray shaded area in Figure 8.6b represents the p-value. The p-value, equal to 0.004, is less than 0.05.

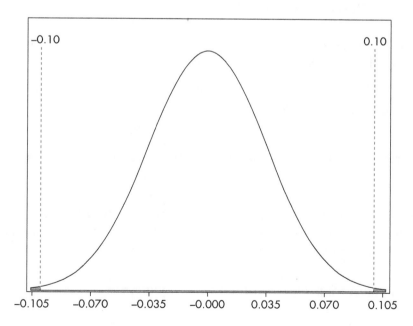

Figure 8.6b: The p-Value for the Hypothesis Test

Data source: http://blogs.sas.com/content/sastraining/2014/06/10/producing-normal-density-plots-with-shading/

Step IV – Conclusion

We reject the null hypothesis in favor of the alternative hypothesis, which states that the population proportion difference is not equal to zero. The p-value of 0.004 is strong statistical evidence that the proportion of men and women in the population who will vote for a particular candidate differ, but it is the confidence interval that gives us a range of plausible values for that difference.

The 95% confidence interval for the population proportion difference (calculated in **Example 7.7**), states that its value lies somewhere between 0.03 and 0.17. Again, we can see the duality between the result of the hypothesis test and the range of plausible values within the confidence interval. The 95% confidence interval does not include zero, and thus the null hypothesis was rejected in favor of the two-sided alternative, at the 0.05 level of significance.

In Chapter 7, when discussing confidence intervals, we looked at a survey from the Pew Research Center titled "Political Polarization and Media Habits." In particular, we looked at two questions comparing Americans considered consistently conservative (CC) and consistently liberal (CL). The findings were as follows:

1. Consistent conservatives see more Facebook posts in line with their views than consistent liberals, 47% versus 32%, respectively.
2. Consistent liberals more likely to block others because of politics than consistent conservatives, 44% versus 32%, respectively.

We will now complete a hypothesis test comparing the proportions in each group for both questions.

1. Is there a difference in the population proportion of consistent conservatives and the population proportion of consistent liberals (CC minus CL) who view Facebook posts in line with their views?

For the first question, the null and alternative hypotheses are as follows:

Step I – Hypotheses

Null Hypothesis: population proportion difference equal to zero ($H_0: p_{CC} - p_{CL} = 0$)
Alternative Hypothesis: population proportion difference not equal to zero ($H_A: p_{CC} - p_{CL} \neq 0$)

Step II – The Model

Using the previously calculated standard error of 0.044 (see appendix for calculation), we can calculate the range of sample proportion differences we would expect, given the population proportion difference is equal to zero:

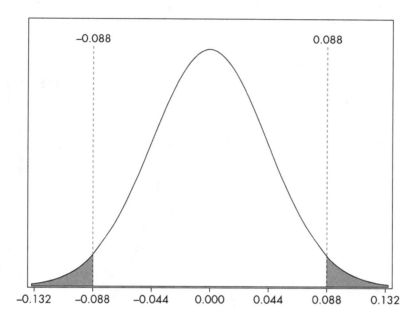

Figure 8.7a: The Model of Possible Sample Proportion Differences (Based on a Total Sample Size of 953–309 CC and 644 CL)

Data source: http://blogs.sas.com/content/sastraining/2014/06/10/producing-normal-density-plots-with-shading/

If the population proportion difference is zero, we expect 95% of sample proportion differences to fall between –0.088 and 0.088. The sample proportion difference between consistent conservatives and consistent liberals of 0.15 or 15% (47% minus 32%) lies well beyond these limits.

Step III – Calculations

$$z = \frac{(\hat{p}_{CC} - \hat{p}_{CL}) - (p_{CC} - p_{CL})}{se(\hat{p}_{CC} - \hat{p}_{CL})}$$

test statistic = (sample proportion difference − null value)/standard error of the difference
 = ((0.47 − 0.32) − 0)/0.044
 = 0.15/0.044
 = 3.4

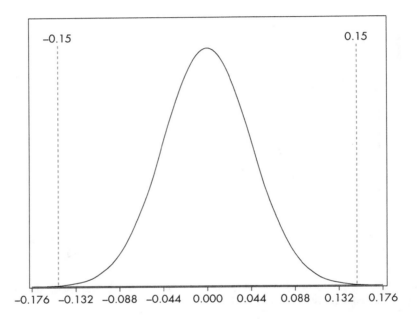

Figure 8.7b: The p-Value for the Hypothesis Test

Data source: http://blogs.sas.com/content/sastraining/2014/06/10/producing-normal-density-plots-with-shading/

The test statistic of 3.4 tells us that the sample proportion difference of 0.15, or 15%, is 3.4 standard errors greater than the the null value of zero. The p-value is equal to 0.0006, which is less than 0.05. The p-value is so small it is difficult to visualize in Figure 8.7b.

Step IV – Conclusion

We reject the null hypothesis in favor of the alternative hypothesis, that the population proportion difference is different from zero. We have found strong statistical evidence that the percentage of consistent conservatives in the population who see Facebook posts in line with their views is different from the percentage of consistent liberals in the population who see Facebook posts in line with their views. According to the confidence interval (calculated in Chapter 7), the proportion (percentage) difference in the population could be anywhere between 0.062 and 0.238 (or between 6.2% and 23.8%). In other words, we have strong statistical evidence that a higher percentage of consistent conservatives than consistent liberals see Facebook posts in line with their views.

2. Is there a difference in the proportion of consistent liberals and the proportion of consistent conservatives (CL minus CC) who block others because of politics?

For the second question of interest, the null and alternative hypotheses are as follows:

Step I – Hypotheses

Null Hypothesis: population proportion difference equal to zero (H_0: $p_{CC} - p_{CC} = 0$)
Alternative Hypothesis: population proportion difference not equal to zero (H_A: $p_{CL} - p_{CC} \neq 0$)

Step II – The Model

Using the same standard error of 0.044 (see appendix for calculation), we can calculate the range of sample proportion differences we would expect, given the population proportion difference is equal to zero:

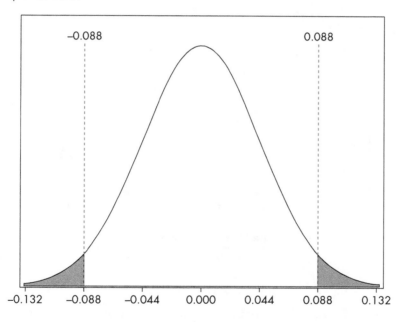

Figure 8.8a: The Model of Possible Sample Proportion Differences (Based on a Total Sample Size of 953–309 CC and 644 CL)

Data source: http://blogs.sas.com/content/sastraining/2014/06/10/producing-normal-density-plots-with-shading/

If the population proportion difference is zero, we expect 95% of sample proportion differences to fall between –0.088 and 0.088, or –8.8% and 8.8%. The sample proportion difference between consistent conservatives and consistent liberals of 0.12 or 12% (44% minus 32%) lies beyond these limits.

Step III – Calculations

$$z = \frac{(\hat{p}_{CL} - \hat{p}_{CC}) - (p_{CL} - p_{CC})}{se(\hat{p}_{CL} - \hat{p}_{CC})}$$

test statistic = (sample proportion difference – null value)/standard error of the difference
= ((0.44 – 0.32) – 0)/0.044
= 0.12/0.044
= 2.73

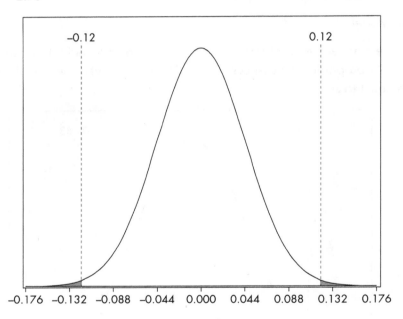

Figure 8.8b: The p-Value for the Hypothesis Test

Data source: http://blogs.sas.com/content/sastraining/2014/06/10/producing-normal-density-plots-with-shading/

The test statistic of 2.73 tells us that the sample proportion difference of 0.12, or 12%, is 2.73 standard errors greater than the null value of zero. The p-value, the gray shaded area in Figure 8.8b, is equal to 0.006, which is less than 0.05.

Step IV – Conclusion

We reject the null hypothesis in favor of the alternative hypothesis, which states that the population proportion difference is statistically significantly different from zero. We have found strong statistical evidence the percentage of consistent liberals in the population who block others because of politics is different from the percentage of consistent conservatives in the population who block others because of politics. According to the confidence interval, the population proportion (percentage) difference could be anywhere between 0.032 and 0.208 (or between 3.2% and 20.8%). In other words, we have strong statistical evidence that a higher percentage of consistent liberals than consistent conservatives block others because of politics. As was the case when discussing the confidence intervals, the necessary assumptions and conditions must hold for the results of the hypothesis tests to be valid.

8.10 ANALYSIS OF 2 X 2 TABLES—CHI-SQUARE TEST

An alternative but equivalent approach (resulting in the same p-value) for comparing proportions (or percentages) is known as a chi-square test of independence. The data is analyzed from the point of view of a two-by-two table. The table contains the counts (or number of outcomes) with particular values of the two categorical variables in our sample data. For example, say we wanted to test whether there is a difference between males and females in how they answer a yes-no question on an important social issue. In other words, are the two categorical variables (gender and answer to a question on social issue) independent or are they related? In other words, is there an association between gender and how people answer the question?

Let's say we were to select a (hypothetical) sample of 100 males and 100 females and ask them a yes-no question on a social issue. Our alternative (research) hypothesis is that the proportion of males and females answering yes to the question differs in the population. In other words, that the two categorical variables are not independent. We decide to conduct a chi-square test of independence to test whether there is statistical evidence in our data that the two categorical variables are not independent.

The null and alternative hypothesis are as follows:

Null Hypotheses: opinion on social issue is independent of gender
Alternative Hypothesis: opinion on social issue is not independent of gender

We begin our reasoning from the point of view of assuming the null hypothesis is true. If the null hypothesis is true, the following two-by-two table shows a breakdown of the counts we would expect to see in our sample of data:

Table 8.2: Expected Counts

	Gender		
Social Issue	Male	Female	Total
Yes	60	60	120
No	40	40	80
Total	100	100	200

The two-by-two refers to the two columns for gender—male and female—and the two rows for a social issue—yes and no in our table. In our (hypothetical) sample of 100 males and 100 females, 120 answered yes and 80 answered no. If the two categorical variables are independent (null hypothesis is true), we would expect 60 yes answers and 40 no answers for both males and females. In other words, we would expect an equal proportion (or percentage) of yes and no answers for both males and females.

As before, we are reasoning with sampling variation in order to make a statistical decision. Our observed counts (from our sample data) could deviate somewhat from our expected counts (under the null hypothesis), and the null hypothesis could still be true. However, when our observed counts deviate so far from the expected counts that they are unlikely to occur by chance (given the null hypothesis is true), we will reject the null hypothesis in favor of the alternative. We have found statistical evidence that the proportion of males and females who answer yes (or no) to the question differ in the population.

We calculate what is known as the chi-square test statistic and use it to calculate the p-value. The chi-square test statistic is calculated as follows:

$$X^2 = \sum_{i=1}^{\text{all cells}} \frac{(O-E)^2}{E}$$

Chi-Square Test Statistic = Sum of (observed counts − expected counts)^2/expected counts

for each of the four cells in our two-by-two table.

In our hypothetical example, the expected counts were easy to determine intuitively because we had easy numbers to work with: 100 men and 100 women. In general, how we calculate the expected count for each cell is as follows:

Expected Count = (Row Total × Column Total)/Table Total

For the first cell in our example (Gender: Male, Social Issue: Yes), the expected count is calculated as follows:

$$\text{Expected Count} = (120 \times 100)/200 = 60$$

The core assumption that the data is a random sample resulting in independent measurements remains the same. For this approach for comparing proportions, the sample size condition states that the expected cell counts all need to be greater than 5 for the test to be valid. If they are not, the test should not be conducted.

If our observed counts turn out to be exactly equal to our expected counts, the chi-square statistic will be equal to zero. In other words, we have zero statistical evidence in favor of the alternative hypothesis. However, the further our observed counts are from our expected counts, the larger (positive value) the chi-square test statistic will be. The larger the test statistic, the smaller the p-value. When analyzing a two-by-two table, the p-value will be less than 0.05 when the chi-square test statistic is greater than 3.84.

As mentioned previously, in our final chapter titled Integrity in Research, we will critique the results of a study reported on extensively in the media titled "Power Posing: Brief Nonverbal Displays Affect Neuroendocrine Levels and Risk Tolerance." The idea of the research was very simple; a person can instantly feel more powerful by simply taking on what the researcher termed a "high-power pose" for as little as two minutes. The researcher, Dr. Amy Cuddy, found what appears to be strong evidence that if you take two minutes to pretend you are Superman or Wonder Woman, you are more likely to take risks.

For her primary analysis on risk-taking, Dr. Cuddy found that 86.36% of the high-power posers were willing to take a gambling risk compared to only 60% of the low-power posers, a difference of 26.36%. She compared the percentages using a chi-square test of independence. In other words, the researcher was looking to see if the type of power-pose a person took on was independent of whether or not the person was willing to take a risk. The analysis resulted in a chi-square test statistic equal to 3.86 with the p-value reported as "<0.05." Since the chi-square test statistic was only slightly greater than 3.84, the actual p-value was equal to 0.0495 just under the threshold of 0.05 for statistical significance. We will discuss how this statistically significant result may have come about in our final chapter.

The analysis of two-by-two tables is an easy to understand and much used alternative for comparing proportions. Another much used statistic that can be calculated from the a two-by-two table is known as **relative risk**. **Risk** is a term used in epidemiology (the study and analysis of health outcomes and diseases in populations) defined as the probability that an event (such as heart attack or death) will occur. Relative risk is a value that compares one group's risk of a disease or outcome relative to another group. A hypothetical example is presented in Table 8.2.

Table 8.2

Group	Heart Attack	No Heart Attack	Total
High Cholesterol	20	80	100
Low Cholesterol	15	135	150
Total	35	215	250

Table 8.2 presents the health outcomes (heart attack or no heart attack) for two groups of patients: one group with high cholesterol (HC) and the other group with low cholesterol (LC). There are 100 patients in the high cholesterol group, of which 20 experienced a heart attack. Therefore, the risk of heart attack for this group of patients is 20/100 equal to 0.20. There are 150 patients in the low cholesterol group, of which 15 experienced a heart attack. Therefore, the risk of heart attack for this group of patients is 15/150 equal to 0.10. We can calculate the relative risk of getting a heart attack as follows:

$$\text{Relative Risk} = \text{Risk for HC patients}/\text{Risk for LC patients}$$
$$= 0.20/0.10$$
$$= 2$$

What the relative risk of 2 means is that the risk of getting a heart attack for the high cholesterol group is twice the risk it is for the low cholesterol group. We could have calculated the relative risk as follows:

$$\text{Relative Risk} = \text{Risk for LC patients}/\text{Risk for HC patients}$$
$$= 0.10/0.20$$
$$= 0.5$$

In this case, the relative risk means that the risk of getting a heart attack for the low cholesterol group is half the risk it is for the high cholesterol group. Both ways of calculating relative risk are valid. However, how the researcher calculates the relative risk should be consistent for all health outcomes of interest. In our final chapter (Integrity in Research), we look at the case of Vioxx, a pain drug taken off the market in 2004 due to the fact it was causing heart attacks in a small but significant proportion of the population. We will look at the results of a large study where the risk of heart attacks on Vioxx first became apparent. We will question why the pharmaceutical company did not consistently calculate relative risk for all health outcomes of interest when presenting the results in the journal article. As you can see from the above example, a relative risk of 2 can seem more alarming than a relative risk of 0.5 even though they mean the same thing. We will discuss how the inconsistencies in the calculation of relative risk may have helped the pharmaceutical company obscure the risk of heart attack on Vioxx.

There have been a number of news articles in the media reporting on the link between eating processed meats and breast cancer risk. Two of the news articles, one in the *New York Times* titled "Eating Processed Meats Tied to Breast Cancer Risk" and one on the *BBC News* website titled "Processed meat 'linked to breast cancer'" discuss the link.

Both articles reported that eating processed meats increases the risk of breast cancer by 9%. However, the *BBC News* article went on to discuss how reliable the findings were and what the 9% increased risk actually meant. The *New York Times* article did not critique the findings or explain what they mean in any way. We will go to the source of the research, the journal article.

The results of the research were reported in the *International Journal of Cancer* in an article titled "Consumption of red and processed meat and breast cancer incidence: A systematic review and meta-analysis of prospective studies." The study was a meta-analysis that combined the results of a number of prospective studies, similar to the alcohol study we critiqued in Chapter 2. Regarding the link between processed meats and breast cancer risk, the journal article stated the following:

> Among 15 studies that evaluated the association between processed meat and overall breast cancer, the risk estimate comparing the highest vs. the lowest category was 1.09 (95% CI: 1.03, 1.16)

Reading through the journal article, nowhere is it explicitly stated how the lowest and highest categories of processed meat consumption were defined. The *New York Times* reporter stated that high consumption was an average of 25–30 grams per day while low consumption was an average 0–2 grams per day. It is not clear where the reporter got this information. Perhaps the news article was simply based on a conversation the reporter had with one of the researchers on the phone or through email. As already stated, the *BBC News* reporter took a closer look at the results of the study. He points out that across the 15 studies there

NEWS ARTICLE: EATING PROCESSED MEATS TIED TO BREAST CANCER RISK

Web Link: https://www.nytimes.com/2018/10/03/well/eat/eating-processed-meats-tied-to-breast-cancer-risk.html?module=inline

Search Term: Eating Processed Meats Tied to Breast Cancer Risk

NEWS ARTICLE: PROCESSED MEAT 'LINKED TO BREAST CANCER'

Web Link: https://www.bbc.co.uk/news/amp/health-45720970

Search Term: Processed meat "linked to breast cancer"

RELATED JOURNAL ARTICLE: CONSUMPTION OF RED AND PROCESSED MEAT AND BREAST CANCER INCIDENCE: A SYSTEMATIC REVIEW AND META-ANALYSIS OF PROSPECTIVE STUDIES

Web Link: https://onlinelibrary.wiley.com/doi/abs/10.1002/ijc.31848

Search Term: Consumption of red and processed meat and breast cancer incidence

were highly varying definitions for what the highest category of consumption meant, ranging from 9 grams a day to much higher values.

The relative risk (highest relative to lowest) was equal to 1.09. In other words, there was a 9% increased risk of breast cancer for individuals in the highest category (of processed meat consumption) compared to the lowest category. To understand what this actually means, we would need to know what is the actual risk of getting breast cancer for those individuals in the lowest category. It is very difficult to know what that value is. As an estimate, the *BBC News* reporter stated that (in the UK) over a lifetime, 14 out of every 100 people get breast cancer, an overall lifetime risk of 14%. If the 9% increased risk of breast cancer (from high consumption of processed meats) is an accurate estimate of the true increased risk in the population, then the overall risk would increase to approximately 15% (14 + 14 × 0.09 = 15.26). This explanation of what the results mean is far less alarming to the reader than simply reporting the increased risk is 9%.

As was the case with the alcohol study, the *BBC News* reporter also points out that there are many other factors that differ between individuals in the study that may affect their chances of getting breast cancer. Our focus here was to simply discuss what the relative risk means. However, it would be an interesting exercise to critique the overall quality of the research as we did with the alcohol study.

8.11 TYPES OF ERROR AND THE POWER OF THE TEST

In this section, we will discuss what the chances are that you choose the wrong hypothesis, known as Type I error and **Type II error,** when conducting a hypothesis test. We will also discuss the power of the test: the chances you choose your alternative (research) hypothesis when it is actually true. We will reason through these concepts using the coin-toss experiment and calculate probabilities for each. We will learn that at the beginning of a study, a researcher should calculate the sample size necessary to ensure a sufficiently high-power probability. In practical application, statistical software will take care of these calculations. What we need to focus on is the meaning of these concepts, how they relate to each other, to effect size and to sample size.

We will learn that the larger the true (population) effect size, the greater the power (probability) your test will have to accept a true alternative. We will learn that the smaller the true effect size, the lower the power (probability) to accept a true alternative. When the true effect size is small, increasing the sample size will increase the power of the test to reject the null hypothesis in favor of a true alternative (research) hypothesis. The sample size is the fuel underneath the hood giving your test greater power.

In any hypothesis test that we conduct, we must make one of two choices:

Choice A: Rejecting the Null Hypothesis in Favor of the Alternative
Choice B: Failing to Reject the Null Hypothesis

We are making a decision based in uncertainty due to sampling variation, so there is always a chance that we made the wrong decision. The two potential errors we can make are as follows:

Type I Error: Rejecting the null hypothesis in favor of the alternative when it is true
Type II Error: Failing to reject the null hypothesis when it is false

At the beginning and end of our research, we need to think about the probability of committing an error. When we are setting up our hypothesis test, we decide on a level of significance that is acceptable. As we have discussed, a 0.05 level of significance is the most commonly used borderline value for making statistical decisions using hypothesis testing.

The 0.05 (or 5%) level of significance is our type I error probability (also known as the alpha level) we choose at the beginning of our hypothesis test. It means that we are willing to accept a 5% chance of rejecting the null hypothesis, when the null hypothesis is true. In other words, we are willing to accept a 5% chance of committing a type I error. With a type I error rate of 5%, 1 out of every 20 studies (on average) would result in a statistically significant result when the null hypothesis is actually true.

Once we have completed our hypothesis test, we can think of our p-value as our observed level of significance (or our observed type I error probability). For example, if our p-value was equal to 0.001, we could have begun our test with a level of significance equal to 0.01 and still rejected the null hypothesis, since our p-value is less than 0.01. If we started with a level of significance equal to 0.002, we would still have rejected the null hypothesis, since our p-value is less than 0.002. Therefore, we can think of our p-value as the lowest level of significance for which we would reject the null hypothesis in favor of the alternative hypothesis.

A p-value of 0.001 means that by rejecting the null hypothesis, there is only a 1 in 1,000 chance that we made the wrong decision of rejecting the null hypothesis when it is true, a type I error. The smaller the p-value, the less likely we've made a type I error, and the stronger the statistical evidence in favor of the alternative hypothesis.

Determining the probability for whether we made a type II error (also known as the beta level) is somewhat more challenging. Let's return to the coin-toss experiment from the beginning of this chapter to help us understand why. The null and alternative hypotheses were as follows:

Null Hypothesis: population proportion of heads is equal to 0.5 (H_0: p = 0.5)
Alternative Hypothesis: population proportion of heads is not equal to 0.5 (H_A: p ≠ 0.5)

We can think of the null versus the alternative in this example as analogous to no treatment effect versus a treatment effect in the population. If we fail to reject the null hypothesis, we may have committed a type II error. The coin may be biased toward heads or tails (a treatment effect) but we have failed to find statistical evidence in favor of that conclusion.

At the end of the hypothesis test, if we fail to reject the null hypothesis that the coin is fair, we would like to be able to calculate the probability that we made a type II error. However, there is really no way for us to calculate such a value. We would have to know how biased the coin is (for example, biased toward heads 60% of the time or a 10% effect size) to determine this probability. However, if we knew that, then there would be no need to conduct the test. What we can do, before we conduct the hypothesis test, is calculate type II error probabilities based on different possible values for how biased the coin might be. In other words, by choosing different alternative values for the population parameter (or alternative effect sizes), we can calculate the probability of failing to detect the alternative is true for each alternative value chosen.

In our coin-toss experiment, we state the value for the population proportion in the null hypothesis and decide upon the type I error rate or level of significance (usually 0.05) we are willing to accept. This enables us to determine the sample proportion values (or borderline values) for rejecting the null hypothesis. If our sample proportion of heads is more extreme than the borderline values, we will reject the null hypothesis in favor of the alternative that the coin is biased.

The more biased a coin is toward heads or tails, the smaller the probability of committing a type II error will be, and the more likely our hypothesis test will detect the bias in the coin. In other words, the more biased a coin is, the more likely we will obtain a sample proportion of heads that deviates far enough from 0.50 (or 50%) for us to reject the null hypothesis.

For example, if the coin is biased toward heads 80% of the time (or a 30% effect size), what is the probability of failing to reject the null hypothesis and committing a type II error?

---

* Technically, when the population proportion of heads is 0.80 as in Figure 8.9b, the standard error is equal to 0.04. See appendix for formula calculation.

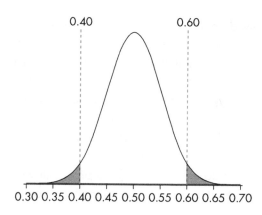 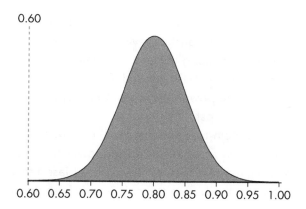

Figure 8.9a: Coin Is Fair (100 Coin Tosses)
Figure 8.9b: Coin Is Biased toward Heads 80% of Coin Tosses (100 Coin Tosses)

Data source: http://blogs.sas.com/content/sastraining/2014/06/10/producing-normal-density-plots-with-shading/

When we toss a coin 100 times, the standard error of the sample proportion of heads is equal to 0.05. If the coin is fair, we expect 95% of sample proportions to be between 0.40 and 0.60 (or 40% and 60%) or two standard errors from 0.5. If the coin is biased toward heads 80% of the time, we would expect 95% of sample proportions to be between 0.70 and 0.90 (or 70% and 90%). *There is no overlap between these two ranges of possible sample proportions.

If the coin is biased toward heads 80% of the time, the probability of getting a sample proportion of heads greater than or equal to 0.60 (or 60%)—and therefore accepting a true alternative hypothesis—is very high or close to one. In other words, our sample proportion of heads comes from the sampling distribution shown in Figure 8.9b. It is almost certain we will obtain a value greater than 0.60 (the gray shaded area under the curve in this figure). Since the probability of accepting a true alternative is very high or close to one, we can say the probability of failing to reject the (false) null hypothesis (a type II error) is very low or close to zero.

The probability of accepting a true alternative is known as the power of the test. The power of the test and type II error probabilities are known as complement probabilities. For example, if the probability it is going to rain tomorrow is 0.3, the probability it is not going to rain is 0.7. These probabilities are considered the complement of each other.

Another way of stating a type II error is that it is the probability of failing to accept a true alternative. The power of the test is the probability of accepting a true alternative. When the alternative is true, there are only two possible outcomes: accepting or failing to accept a true alternative. They are complements of each other, and their combined probabilities must add up to one. The larger the power of the test probability, the smaller the type II error probability, and vice versa.

In any study, if the alternative is true, (say, a new drug for cholesterol is effective), our goal is to find statistical evidence in favor of that conclusion. If we spend a lot of time and money conducting research and fail to show a drug is effective (when it actually is), then an opportunity has been missed. As already stated, at the beginning of our research, we need to calculate the power of our test to accept the alternative hypothesis based on particular effect size(s) of interest.

In our coin-toss experiment, the more biased the coin, the higher the probability of accepting the true alternative and the lower the probability of failing to accept a true alternative. As we already discussed, if a coin is biased toward heads 80% of the time, the probability of detecting the bias in the coin is very high. In other words, the test will have very high power probability.

It is very important that your experiment has an acceptable power probability. If a new drug is effective, but the hypothesis test only has a 20% (power) probability of detecting such an effect, then the test has very low power. There are two reasons why a hypothesis test may have low power.

The first reason is that the true effect size might be very small, making it very hard to detect using hypothesis testing. In our coin-toss experiment of 100 coin-tosses, what if the coin is biased toward heads 51% of the time? We can think of this 1% bias in the coin as a 1% effect size.

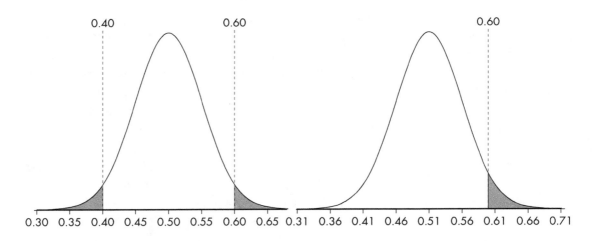

Figure 8.10a: Coin Is Fair (100 Coin Tosses)
Figure 8.10b: Coin Is Biased toward Heads 51% of Tosses (100 Coin Tosses)

Data source: http://blogs.sas.com/content/sastraining/2014/06/10/producing-normal-density-plots-with-shading/

The range of possible sample proportions of heads we expect assuming the coin is fair will be very similar to what we would expect, given the coin is biased toward heads 51% of the time. If the coin is fair, we expect 95% of sample proportions to be between 0.40 and 0.60 (or 40% and 60%). If the coin is biased toward heads 51% of the time, we would expect 95%

of sample proportions to be between 0.41 to 0.61 (or 41% and 61%). The range of possible sample proportions are very similiar.

In order to reject the null hypothesis (that the coin is fair) in favor of the alternative that the coin is biased, we would have to obtain a sample proportion of heads greater than 0.60. Given that the coin is actually biased toward heads 51% of the time (a 1% bias), the probability of getting a sample proportion of heads greater than or equal to 0.60 (or 60%) is equal to 0.054. This probability is represented by the gray shaded area in Figure 8.10b. In other words, the probability of accepting the true alternative, the power of the test, is very low when the truth is very close to the null value (or when the population effect size is very small).

When comparing treatments, say a placebo versus actual treatment, a 1% effect size could mean that the actual treatment is 1% more effective (say 1% more patients respond to the actual treatment) than to the placebo. In hypothesis testing, we reason with sampling variation as a way of determining whether the drug is effective. The further the sample effect size is from the null hypothesized value of zero for the population effect size, the more likely we will accept the true alternative. However, if the population effect size of the drug is very small, it will be very difficult to detect using hypothesis testing. As will become clear from our next discussion, the sample size needs to be very large to detect such small effect sizes.

The second reason a study may have low power is due to sample size. When the sample size is small, there is wide variation in the possible sample statistics around the unknown population parameter.

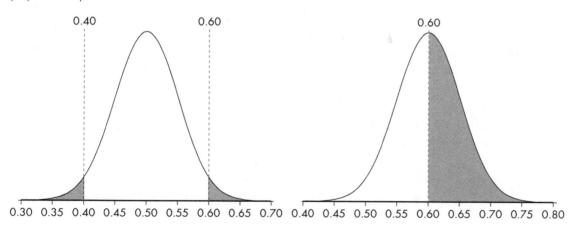

Figure 8.11a: Coin Is Fair (100 Coin Tosses)
Figure 8.11b: Coin Is Biased toward Heads 60% of Tosses (100 Coin Tosses)

Data source: http://blogs.sas.com/content/sastraining/2014/06/10/producing-normal-density-plots-with-shading/

Looking at Figure 8.11a, we expect 95% of sample proportions to be between 0.40 and 0.60 (or 40% and 60%). If the coin is biased toward heads 60% of the time, we would expect 95% of sample proportions to be between 0.50 to 0.70 (or 50% and 70%), as shown in Figure

180 | STATISTICAL THINKING THROUGH MEDIA EXAMPLES

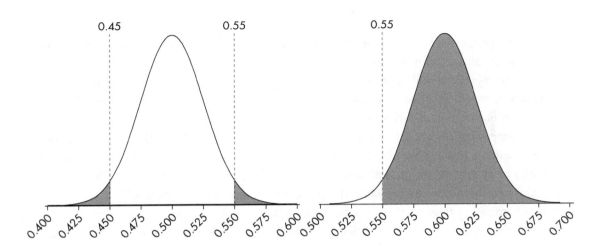

Figure 8.12a: Coin Is Fair (400 Coin Tosses)
Figure 8.12b: Coin Is Biased toward Heads 60% of Tosses (400 Coin Tosses)

Data source: http://blogs.sas.com/content/sastraining/2014/06/10/producing-normal-density-plots-with-shading/

8.11b. There is quite a good deal of overlap between the sample proportions we expect to obtain assuming the coin is fair, compared to when the coin is biased toward heads 60% of the time.

Given that the coin is actually biased toward heads 60% of the time (a 10% effect size), the probability of getting a sample proportion of heads greater than or equal to 0.60 (or 60%) is equal to 0.50. This probability is represented by the gray shaded area in Figure 8.11b. In other words, the probability of accepting the true alternative, the power of the test, is equal to 0.50.

Increasing the sample size means that there will be less variation in the possible sample statistics we obtain. If we toss what we assume is a fair coin 400 times, we expect 95% of possible sample proportions of heads to be between 0.45 and 0.55 (or 45% and 55%). If our sample proportion of heads falls outside this range of values, we would say that we found statistical evidence that the coin is biased.

If we toss a coin 400 times that is actually biased toward heads 60% of the time, we expect 95% of possible sample proportions of heads to be between 0.55 and 0.65 (or 55% and 65%). There is very little overlap between the sample proportions we expect to obtain assuming the coin is fair, compared to the sample proportions we expect assuming the coin is biased toward heads 60% of the time.

Given that the coin is biased toward heads 60% of the time, the probability of getting a sample proportion of heads greater than 0.55 (or 55%) is equal to 0.975. This probability is represented by the gray shaded area in Figure 8.12b. In other words, the probability of accepting the true alternative, the power of the test, is equal to 0.975 and the probability of failing to accept the true alternative, type II error, is equal to 0.025. The power of the test to detect the true alternative is higher in this case because of the larger sample size.

The larger the sample size, the easier it will be to detect a population effect size if one exists. The researcher will decide at the beginning of the study what sample size is required to ensure that there is a high probability of accepting a true alternative based on a particular effect size. The most commonly used value for this probability, the power of the test, is 0.80. In other words, the researcher will calculate a sample size that ensures there is an 80% chance of accepting a true alternative. The smaller the population effect size, the larger the sample size needed to make the correct decision and accept the alternative hypothesis.

The power of the test should be calculated by the researcher and discussed in the journal article. A good example of the calculation of the sample size and associated power was published in the *New England Journal of Medicine* in a journal article titled "Targeting BTK with Ibrutinib in Relapsed or Refractory Mantle-Cell Lymphoma." The researchers were looking at the use of a drug called Ibrutinib for the treatment of mantle-cell lymphoma, a type of cancer of the blood. Patients were enrolled in the study based on whether or not they received prior treatment with an anti-cancer drug called Bortezomib. Initially, the researchers wanted to test whether a certain percentage of the patients in both groups would respond to the drug before moving to Stage 2 of the study.

In the following excerpt from the journal article, the researchers discuss the sample sizes required to ensure sufficient power to detect the specific differences in response rates they were looking for.

> *For the cohort of patients without prior treatment with bortezomib, a two-stage design was planned to test the null hypothesis that the response rate would be 20% or less (i.e., before the investigators could proceed to stage 2 of the study, at least 6 of 25 patients had to have a response). We calculated that a sample of 65 patients would provide 91% power to test a difference in the response rate of 20% versus 40% at a one-sided alpha level of 0.01.*
>
> *For the cohort of patients with prior bortezomib treatment, a two stage design was planned to test the null hypothesis that the response rate would be 15% or less (i.e., before the investigators could proceed to stage 2 of the study, at least 5 of 25 patients had to have a response). We calculated that a sample of 50 patients would provide 80% power to test a difference in the response rate of 15% versus 35% at a one-sided alpha level of 0.01.*

TARGETING BTK WITH IBRUTINIB IN RELAPSED OR REFRACTORY MANTLE-CELL LYMPHOMA

Web Link: http://www.nejm.org/doi/full/10.1056/NEJMoa1306220

Search Term: Targeting BTK with Ibrutinib in Relapsed or Refractory Mantle-Cell Lymphoma

In our final chapter titled Integrity in Research, we will discuss in depth the results of a study reported on extensively in the media titled "Power Posing: Brief Nonverbal Displays Affect Neuroendocrine Levels and Risk

Tolerance." The idea of the research was very simple; a person can instantly feel more powerful by simply taking on what the researcher termed a "high-power pose" for as little as two minutes. The researcher, Dr. Amy Cuddy, found what appears as strong evidence that if you take a minute to pretend you are Superman or Wonder Woman, you're more likely to take risks.

However, one of the problems with the study was a very small sample size, a total number of 42 participants randomized to take on a high-power pose or a low-power pose. The journal article does not state what the actual power of the test was, the probability of finding a difference in willingness to take a risk (population effect size) between the two groups if one actually exists. However, with such a small sample size, we can assume that the power of the test was low. It turns out that the researcher did find a statistically significant sample effect size. Dr. Cuddy found that 86% of the high-power posers versus 60% of the low-power posers were willing to take a gambling risk—a sample effect size of 26%. We will discuss how this result may have come about in our final chapter.

The controversial results of the study by Dr. Cuddy led some researchers to try and replicate her findings. The results of this study can be found in the journal *Psychological Science* in an article titled "Assessing the Robustness of Power Posing: No Effect on Hormones and Risk Tolerance in a Large Sample of Men and Women." This study used a total sample size of 200 men and women, and the researchers completed a power analysis. Based on this sample size (and the sample effect size found in Cuddy's research), the power of their test to detect a population effect size (like the sample effect size Cuddy found) was 0.99, or 99%.

In other words, their test had a 99% probability of accepting the alternative hypothesis if the population effect size was equal to 26%. As we will discuss in our final chapter, the researchers found no statistical evidence of any real effect in the population.

8.12 CONCLUSION

In this chapter, we learned how to reason through the process of hypothesis testing for making statistical decisions regarding the value of four key population parameters of interest: population mean, population proportion, population mean difference, population proportion difference. We learned how to compare several means using ANOVA and how to compare proportions using a chi-square test. We learned that if our sample statistic deviates too far from what we would expect, given the null hypothesized value is correct, we should reject the null hypothesis in favor of our alternative (research) hypothesis. We learned that we should question the use of a one-sided alternative (over a two-sided alternative) and whether or not a one-sided alternative is appropriate to use in certain circumstances.

We learned that the most commonly used level of significance for making statistical decisions is 0.05 (or 5%), also known as our type I error rate. It means that we are willing to take a 5% chance of rejecting the null hypothesis when it is true. We learned that our type II error rate, the probability of failing to reject the null hypothesis when it is false (or failing to accept a true alternative) depends on what is the true value of the population parameter of interest. Finally, we learned that at the beginning of a study, we should calculate the sample size required to have a high probability of detecting a true alternative, the power of our test.

In our next chapter, we will build on the foundations in statistical thinking we laid down throughout this book to introduce one of the most important techniques in statistics known as linear regression.

8.13 REAL-WORLD EXERCISES

1. Exercise 2 in Chapter 3 asked you to design a survey asking ten questions, both categorical and quantitative. Use the data collected to complete the following:

    a. Compare two groups (example: Males and Females) on their response to a quantitative question by conducting a hypothesis test for the population mean difference.

    b. Compare two groups (example: Males and Females) on their response to a categorical question by constructing a 95% confidence interval for the population proportion difference.

    Before conducting the hypothesis tests, be sure to check the necessary assumptions and conditions for statistical analysis. State clearly what the results of the hypothesis test mean in terms of the original question of interest.

2. Search the Internet for data of interest to you. One good source of data is the US government website www.data.gov. It contains a searchable database containing data on numerous topics, including education, climate, and finance. Look for datasets to download that are in XLS or CSV file formats. Both file formats can be opened in Microsoft Excel. Once you locate a dataset, calculate sample statistics for the variables of interest in the dataset and conduct the hypothesis test described in Question 1.

3. Conduct an experiment to test the assumed value for a population proportion. For example, you could test whether a coin is fair by tossing it 100 times and noting the proportion of heads you got. Or you could test a friend's extrasensory perception ability by having him/her guess the suit for one of four face-down playing cards. Repeat the experiment 100 times, noting each time he/she guessed correctly.

a. Complete a hypothesis test to determine whether you have found evidence in your data that the population proportion is not equal to the assumed null hypothesis value. Be sure to discuss and check whether your data adheres to the necessary assumptions and conditions for valid analysis.

b. Explain in context what the results of your hypothesis test means.

4. Find data of interest to you online. From the dataset of your choosing, select a quantitative variable. Select a second categorical variable from the dataset, for example, Gender. Group the values of the quantitative variable by the values of the categorical variable.

   a. Complete a hypothesis test for the population mean difference across groups. Explain what the results of the hypothesis test mean in context. Be sure to discuss and check whether your data adheres to the necessary assumptions and conditions for valid analysis.

   b. How representative do you think the data is of the population you are extending the results of your analysis to?

5. Collect data on a quantitative variable of interest to you where you can separate the variable values into two groups you wish to compare, for example, male and female.

   a. Complete a hypothesis test for the population mean difference across groups. Explain what the results of the hypothesis test means in context. Be sure to discuss and check whether your data adheres to the necessary assumptions and conditions for valid analysis.

   b. How representative you think the data is of the population you are extending the results of your analysis to?

6. Find a news article that discusses the results of research that compared two groups on a quantitative variable. Find the source of the research, the journal article.

   a. Read the results of the research presented in the abstract or summary of findings in the journal article. Discuss what the results (p-value and confidence interval, if presented) mean in context.

   b. Read the journal article to understand the meaning of the measurement used and how it was collected. Do you think the data adheres to the necessary assumptions and conditions for analysis?

7. Find a journal article that compared two proportions across groups by conducting a comparison of proportions or chi-square hypothesis test.

    a. Read the results of the research presented in the abstract or summary of findings in the journal article. Discuss what the p-value means in context.

    b. Read the journal article to understand the meaning of the categorical measurements used and how they were collected. Do you think the data adheres to the necessary assumptions and conditions for analysis?

8. Find a journal article that conducted a power analysis.

    a. Discuss in your own words and in the context of the research what the power analysis means.

    b. Based on the results of the research, explain in context what type of error the researcher might have made. Discuss any potential consequences for the possible error in his/her decision-making.

CHAPTER 9

BUILDING ON FOUNDATIONS: LINEAR REGRESSION

All models are wrong, but some are useful.

—George Box

9.1 INTRODUCTION

The foundations in statistical thinking (sampling distributions, confidence intervals, and hypothesis testing) that we learned in the last three chapters can be applied to more advanced techniques like linear regression. How confidence intervals are constructed and the four steps of hypothesis testing remain the same. The meaning of the confidence interval and p-value (in terms of the population parameter of interest) remains the same. Only the calculations involved in constructing confidence intervals and obtaining p-values will be different.

In Chapter 4, we summarized single quantitative (measurement) variables like height and weight by creating histograms and calculating sample statistics like the sample mean and the sample standard deviation. In Chapters 7 and 8, we learned how to use the sample statistic to make statistical decisions about the population parameter.

We will now focus on looking at the association or relationship between two quantitative variables like height and weight. In this chapter, we will ask the following questions:

- How do we visualize the relationship between two quantitative variables?
- What sample statistics can we calculate to measure the relationship between two quantitative variables?

KEY TERMS

Scatterplot: A graphical display of data points representing paired values of two quantitative variables for a sample of data

Sample Slope: A sample statistic that represents the angle (positive or negative) of the line we fit to the scatterplot of sample data points

Population Slope: A population parameter that represents the angle (positive or negative) of the line we would (ideally) fit to the population of data points

Deterministic Relationship: A relationship between two variables in which the value of one variable can be completely determined by the value of another variable

Linear Relationship: A relationship between two quantitative variables where the data points in the scatterplot follow a linear trend

(Simple) Linear Regression Model: An attempt to model the relationship between two quantitative variables by fitting a line to the scatterplot of sample data points

Explanatory Variable: The variable used to explain or predict changes in the values of the response variable

Response Variable: The variable (whose variability) we want explained by modeling the relationship between this variable and explanatory variable(s)

Sample Correlation Coefficient: A sample statistic that measures the strength (positive or negative) of the linear relationship between the two quantitative variables in the sample

Population Correlation Coefficient: A population parameter that measures the strength (positive or negative) of the linear relationship between the two quantitative variables in the population

Simple Linear Regression: A linear regression model with a single explanatory variable

Multiple Linear Regression: A linear regression model with more than one explanatory variable

Residual Deviations (Residuals): Vertical distances of an actual value of the response variable to the predicted value of the response variable

Danger of Extrapolation: Making predictions for the response variable beyond the range of values for the explanatory variable

Line of Best Fit: Another name for the linear regression model or line we fit to the scatterplot of data. It refers to the fact that the line is chosen by minimizing the sum of the squared residual deviations

Outlier: A data point with a large residual or an extreme value of the response variable, with the potential to influence the slope of the regression line

Influential Point: An outlier that affects the slope of the regression line. In other words, omitting it would result in a very different sample slope

Fulcrum Point: A point on the regression line representing the mean of both the explanatory and the response variable. The line balances on the fulcrum point like a seesaw, with each of the data points around the fulcrum point determining the slope of the line

R-Square (Coefficient of Determination): The proportion (or percentage) of the variability in the response variable that can be explained by a linear relationship with the explanatory variable(s)

- What do we mean when we say two quantitative variables are correlated?
- What does it mean to fit a line to our sample of measurements?
- How do we choose the line that best summarizes the relationship between two quantitative variables?
- What does the slope of the line (the sample slope) we fit to our sample data mean?
- How do we use the sample slope to make statistical decisions about the population slope?

We will visual the relationship between the variables with a graphical display known as a **scatterplot**. We will summarize the relationship we see in the scatterplot with the **sample correlation coefficient** and the **sample slope**. We will use the sample slope to make statistical decisions about the value of the population parameter known as the **population slope**.

9.2 VISUALIZING RELATIONSHIPS BETWEEN QUANTITATIVE VARIABLES

We will begin with a simple example looking at the relationship between height (in inches) and weight (lbs.) for 19 children. The sample data is from the SAS Institute website. Let's assume that the 19 children are a random sample from the US population of children.

In Figure 9.1, each blue dot represents an individual child's height and weight value. For example, the dot in the lower left-hand corner of the scatterplot represents a child named Joyce with a height of 51.3 inches and a weight of 50.5 lbs. The blue dot in the upper right-hand corner represents a child named Philip with a height of 72 inches and a weight of 150 lbs.

Source: https://support.sas.com/documentation/cdl/en/statug/63033/HTML/default/viewer.htm#statug_reg_sect003.htm

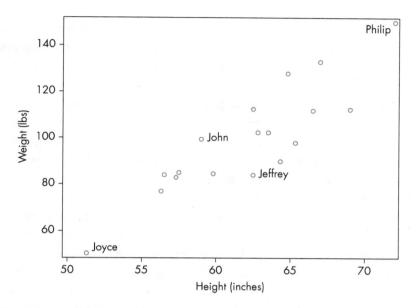

Figure 9.1: Scatterplot of Height versus Weight for 19 Children

Looking at the scatterplot, we can see a positive relationship between the height and weight of the children. In other words, the taller the child, the heavier the child tends to be. We can also see that the relationship looks like what is called a **linear relationship**. The data points look like they are following a linear trend. In other words, we can imagine a line running through the data points from left to right.

However, the relationship between height and weight is not what is called a **deterministic relationship**. A deterministic relationship between two variables is one in which the value of one variable can be completely determined by the value of another variable. For example, Fahrenheit and Celsius, or diameter and the circumference of a circle, have deterministic relationships. The value of one variable can be determined by knowing the value of the other variable. If such a relationship did exist between height and weight (and it were also a linear relationship), all the data points in the scatterplot would fall along a straight line.

As discussed in Chapter 5, in the observational world of data, we don't have complete knowledge and understanding of which factors cause the effects we observe. There are many factors that determine a child's weight. Height and age are just two of those factors. However, there are many other factors, such as genetics and diet, which determine a child's weight. If we were able to measure all the factors that determine a child's weight (and how much each factor affects their weight), then we could predict their weight exactly. However, this is impossible for us to do. For that reason, we can think of the linear relationship we observe between two quantitative variables like height and weight as a statistical (linear) relationship.

What we can do is summarize the relationship between two quantitative variables, like height and weight, by using a statistical model known as a **simple linear regression model**. A simple linear regression model summarizes the relationship by fitting a straight line to the scatterplot. Most, if not all, of the data points in the scatterplot will not fall on the fitted line. This is because of the variation in the weight measurements due to other factors besides height. However, we can use the fitted line to make predictions for the weight of children who are a particular height. In other words, we could say the model is wrong because it does not take into account all the factors (or variables) besides height that affect (or are related to) a person's weight. However, the model is useful for helping us better understand the relationship between height and weight.

When we fit a linear regression model to our scatterplot, we call the variable we place on the x-axis (horizontal) of our scatterplot the **explanatory variable** and the variable on the y-axis (vertical) the **response variable**. They are also known as the independent variable and the dependent variable respectively. The line we fit to the scatterplot is used to determine how much of the variation in the response variable can be explained by its linear relationship with the explanatory variable. In our example of 19 children, height is our explanatory variable and weight is our response variable. We are using the variable height to try and explain (at least some) of the variability in weight measurements. In section 9.4, we will fit a line to the data summarizing the linear relationship between these two variables. We will begin by summarizing the linear relationship with a statistic known as the sample correlation coefficient.

9.3 CORRELATION VERSUS CAUSATION

The sample correlation coefficient measures the strength of the linear relationship between our two quantitative variables, height and weight. A relationship between two quantitative variables is considered linear if the data points in the scatterplot tend to follow a linear trend, across all values of the explanatory variable. If we can imagine a line running through the middle of our data points in the scatterplot from leftmost to rightmost points, then we can consider the relationship to be linear.

The sample correlation coefficient can have a value from –1 to +1. A correlation coefficient of +1 means that there is a positive deterministic (linear) relationship between the two quantitative variables. Large values of the explanatory variable correspond with larger values of the response variable (and vice versa), and all data points fall along a straight line with a positive slope.

A correlation coefficient of –1 means that there is a negative deterministic (linear) relationship between the two quantitative variables. Large values of the explanatory variable coincide with smaller values of the response variable (and vice versa), and all data points fall along a straight line with a negative slope.

A correlation coefficient of zero means there is no linear relationship between the two quantitative variables. The closer the correlation coefficient is to +1 or –1, the stronger the linear relationship (positive or negative) between the two quantitative variables. The following table is a breakdown of what is considered weak, moderate, or strong correlations:

Table 9.1

Weak	Moderate	Strong
–0.5 to 0.5	–0.5 to –0.8 or 0.5 to 0.8	–0.8 to –1 or 0.8 to 1

How we calculate the correlation coefficient is beyond the scope of this book. However, the key point to understand is that the correlation coefficient is a measure of the strength of a linear relationship between two quantitative variables. When visualizing this relationship in a scatterplot, the closer the data points arrange themselves around a straight line (with positive or negative slope), the closer the correlation coefficient will be to +1 or –1. In our height and weight example (see Figure 9.1), the correlation coefficient is equal to 0.88. There is a strong positive linear relationship between height and weight in our sample of 19 children.

In the research on college football playing we discussed in previous chapters, the researchers also looked at the relationship between the number of years of playing experience and reaction times. The correlation coefficient was equal to –0.43, considered a weak negative correlation. In other words, the data indicate that the more years of playing experience a football player has, the slower his reaction times, but the relationship is not very strong. This is what we would expect. Reaction times will vary quite a bit from player to player for any given number of years of playing experience. Reaction time may be slower for some players over time, but for other players, reaction times might actually increase with the added years of experience. There maybe other factors that also affect reaction times. This means that a lot of the variation in reaction times can't be explained by a linear relationship with years of playing experience alone.

Whatever the value of the correlation coefficient between two quantitative variables, we have to understand that correlation does not mean causation. A good (and amusing example) was discussed in a news article by Reuters.

Eat Chocolate, Win the Nobel Prize

Of all the chocolate research out there, the most unabashed tribute to the "dark gold" has to be a study just published in one of the world's most prestigious medical journals.

Drum roll, please: The higher a country's chocolate consumption, the more Nobel laureates it spawns per capita, according to findings released today in the New England Journal of Medicine.

The news article was based on a piece of research presented in the *New England Journal of Medicine* titled "Chocolate Consumption, Cognitive Function, and Nobel Laureates." The researchers examined the relationship between Chocolate Consumption (kg/yr/capita) and Nobel Laureates per 10 Million Population. They found the correlation to be equal to 0.791 with a p-value less than 0.0001. All the p-value is saying is that the sample correlation of 0.791 is statistical evidence that the population correlation is not zero.

In the world of observational data, we will find many variables that are correlated, but that does not mean there is a causal connection. Eating more chocolate is not a major factor in determining the number of Nobel laureates. As the news article points, one common factor driving this relationship is wealth. The wealthier a country, the better its research institutions. The wealthier a country, the better the range and quality of its chocolate, and the greater the amount of extra disposable income for purchasing it. The fact that the Nobel Prizes are awarded by Sweden, a European country, could also be a factor driving the relationship in the data. The countries with the highest number of Nobel laureates and the largest amount of chocolate consumption are Switzerland, Sweden, Denmark, Austria, and Norway, all European countries. Also, these countries have great-tasting chocolate.

The point is that the world of observational data is often very chaotic and complex. If we look hard enough, we will find a relationship between variables, but it does mean that it is a causal relationship. The meaning of causality is the change in the value of one variable will result in the change of the value of another variable. As we discussed in Chapter 1, the only way we can determine if there is a cause-and-effect relationship between two variables is through the use of randomized experiments.

However, you may be asking yourself the question: Is it possible to establish causation in situations where we can't conduct a randomized experiment? In other words, is it ever possible to conclude that there is a cause and effect relationship between two quantitative variables from a correlation based on observational data?

We already answered this question when critiquing the alcohol study in Chapter 2. No, we can't make causal conclusions from observational studies. However, modeling the relationships between variables in observational studies can be a useful aid in helping us come to well-reasoned conclusions about how the world works. We will illustrate this by looking at research into an event that is happening with too much regularity in the United States: public mass shootings. We will look at a study examining the relationship between the (possible) explanatory variable, firearm availability, and the response variable, the number of public mass shootings across countries. By controlling for certain factors across countries and

EAT CHOCOLATE, WIN THE NOBEL PRIZE

Web Link: http://www.reuters.com/article/us-eat-chocolate-win-the-nobel-prize-idUSBRE8991MS20121010

Search Term: Eat Chocolate, win the Nobel Prize

CHOCOLATE CONSUMPTION, COGNITIVE FUNCTION, AND NOBEL LAUREATES

Web Link: http://www.nejm.org/doi/full/10.1056/nejmon1211064

Search Term: Chocolate Consumption, Cognitive Function, and Nobel Laureates

eliminating other possible explanatory variables as explanations for the differences in the response variable across countries, we will see more clearly the relationship between these two variables.

Public mass shootings occur so often in the United States, particularly in high schools, that many of them are no longer headline news. These sorts of attacks occur far more frequently in the United States than anywhere else in the world. The question researchers are asking is why; what are the factors that differ between the United States and the rest of the world that are resulting in such a high proportion of public mass shootings? An article in the *New York Times* titled "What Explains U.S. Mass Shootings? International Comparisons Suggest an Answer" discusses the results of research that tried to find an answer to this question.

NEWS ARTICLE: WHAT EXPLAINS U.S. MASS SHOOTINGS? INTERNATIONAL COMPARISONS SUGGEST AN ANSWER

Web Link: https://www.nytimes.com/2017/11/07/world/americas/mass-shootings-us-international.html

Search Term: What Explains U.S. Mass Shootings?

> *When the world looks at the United States, it sees a land of exceptions: a time-tested if noisy democracy, a crusader in foreign policy, an exporter of beloved music and film.*

> *But there is one quirk that consistently puzzles America's fans and critics alike. Why, they ask, does it experience so many mass shootings?*

> *Perhaps, some speculate, it is because American society is unusually violent. Or its racial divisions have frayed the bonds of society. Or its citizens lack proper mental care under a health care system that draws frequent derision abroad.*

> *These explanations share one thing in common: Though seemingly sensible, all have been debunked by research on shootings elsewhere in the world. Instead, an ever-growing body of research consistently reaches the same conclusion.*

> *The only variable that can explain the high rate of mass shootings in America is its astronomical number of guns.*

The research discussed in the article found a strong positive correlation between the number of guns in a country and the number of public mass shootings. A public mass shooting was defined as one where at least four people were killed. The United States has far more guns than any other country, with 88.8 firearms per 100 people, and almost 30 mass shootings per 100 million people from 1966 to 2012. Only Yemen had more public mass shootings over the same time period: 40 public mass shootings per 100 million people with 54.8 firearms per 100 people. The research was conducted by Professor Adam Lankford at the University of Alabama,

JOURNAL ARTICLE: PUBLIC MASS SHOOTERS AND FIREARMS: A CROSS-NATIONAL STUDY OF 171 COUNTRIES

Web Link: https://www.ncbi.nlm.nih.gov/pubmed/26822013

Search Term: Public Mass Shooters and Firearms: A Cross-National Study of 171 Countries presented in the journal *Violence and Victims*, in an article titled "Public Mass Shooters and Firearms: A Cross-National Study of 171 Countries."

So, does this mean that there is a possible causal relationship between these two quantitative variables? In other words, does having more guns in a country lead to or, in some way, cause a greater number of public mass shootings?

Before completing his analysis, Dr. Lankford discusses how mass murders occur in other countries with stricter gun laws than the United States. In China, mass murderers usually have to resort to using knives or blunt instruments. He also discusses the case of Sebastian Boss, a German student who, in 2005, posted online his frustration of not being able to get the firearms he needed. On eventually obtaining the guns he needed 17 months later, he carried out a mass shooting at a school in Emsdetten, Germany. After completing this discussion, Dr. Lankford states that:

> It therefore seems at least possible that cross-national differences in firearms availability may help to explain cross-national differences in rates of public mass shooters. For now, that remains unknown. This study will be the first to provide empirical data on the subject and thus inform the public debate.

In Section 9.6, we will discuss what is known as **multiple linear regression**. It is a very powerful statistical technique for analyzing the complex relationship between variables. It enables us to control for, and take into account, other factors (besides the primary explanatory variable of interest) that may explain (at least some of) the variability in the response variable. In this study, if the relationship between the number of guns and the number of mass shootings across countries remains strong after taking other (potentially) important factors into account, it strengthens the case for a possible causal relationship between these two variables.

The researcher completed a regression analysis across 171 countries, taking into account two other (potential) explanatory variables—homicide and suicide rates. In other words, are the number of homicides and/or suicides in a country any way related to the number of public mass shootings?

While controlling for population size, gender ratios, and percent of urbanization, Dr. Lankford did not find a statistically significant relationship between the number of public mass shootings in a country and its homicide or suicide rates. In other words, higher (or lower) homicide or suicide rates across countries were not associated with a higher (or lower) number of public mass shootings across countries. However, Dr. Lankford did find that the relationship between the number of public mass shootings in a country and the number of guns was statistically significant with (or without) the homicide and suicide variables in the regression model. Dr. Adam Lankford explains his reasoning as to why he thinks these results make sense:

> Public mass shooters often plan their attacks in advance, but other murderers are much more likely to commit crimes of passion or escalation. And although public mass shooters often kill random

strangers or bystanders, most other murderers do not target strangers unless there is something tangible to be gained, such as stolen goods or money.

National suicide rates also failed to explain the global distribution of public mass shooters. In retrospect, this also seems to make sense. Although mental illness can exacerbate many personal problems, the vast majority of people who are mentally ill and suicidal are nonviolent.

In his conclusion, Dr. Lankford states that the most obvious step (but the most politically challenging) would be to reduce firearm availability and points to evidence in Australia where this approach seems to have been effective. After a horrific mass shooting at a New Zealand mosque in March 2019, the government there moved very swiftly to do the same. Dr. Lankford suggests that further research should be conducted to better understand the characteristics of the public mass shooter and why it is such an American problem. He suggests that in a society that puts such a high premium on fame, the need for fame could also be a motivating factor in the mind of a potential public mass shooter in the United States. In fact, one of the high-school students who committed a mass shooting at his high-school in 2018, stated in a video recording taken before the shooting: "When you see me on the news, you'll all know who I am." As Dr. Lankford points out, the media needs to be conscientious of this fact when reporting on public mass shootings.

The pressure on the American teenager to become famous is certainly stronger than in any other country. However, there are other factors that are putting pressure on the American teenager and affecting his/her mental health. The *New York Times* article "Why Are More American Teenagers Than Ever Suffering from Severe Anxiety?" discusses the reasons by focusing on one student in particular:

> The disintegration of Jake's life took him by surprise. It happened early in his junior year of high school, while he was taking three Advanced Placement classes, running on his school's cross-country team and traveling to Model United Nations conferences. It was a lot to handle, but Jake—the likable, hard-working oldest sibling in a suburban North Carolina family—was the kind of teenager who handled things. Though he was not prone to boastfulness, the fact was he had never really failed at anything.
>
> Not coincidentally, failure was one of Jake's biggest fears. He worried about it privately; maybe he couldn't keep up with his peers, maybe he wouldn't succeed in life. The relentless drive to avoid such a fate seemed to come from deep inside him. He considered it a strength.

ARTICLE: NEW ZEALAND IS MOVING TO BAN ASSAULT WEAPONS. WHY CAN'T WE?

Web Link: https://www.cnn.com/2019/03/22/politics/new-zealand-assault-weapons/index.html

Search Term: New Zealand is moving to ban assault weapons

NEWS ARTICLE: WHY ARE MORE AMERICAN TEENAGERS THAN EVER SUFFERING FROM SEVERE ANXIETY?

Web Link: https://www.nytimes.com/2017/10/11/magazine/why-are-more-american-teenagers-than-ever-suffering-from-severe-anxiety.html

Search Term: Why Are More American Teenagers Than Ever Suffering from Severe Anxiety?

The article points to evidence suggesting that heavy use of social media and the increasing pressure on teenagers to get into the best colleges are factors that are also affecting their mental health. Ironically, social media exacerbates feelings of alienation, isolation, and low self-esteem due to cyber bullying and the incessant need to compare oneself to others.

Gun control is a divisive and emotional issue in the United States. Every time another mass shooting makes the headlines, so-called authority figures stir the emotions of people on both sides of the debate, clouding their reasoning on the issue. As discussed in Chapter 2, we should always question the quality of the research put forth by the researchers. We should also question their integrity, the topic of our next and final chapter. However, we also have to trust that there are many good researchers who are dedicated to pursuing the truth no matter what their personal beliefs or opinions are on the issue.

This example illustrates that by controlling for certain factors and including other potential explanatory variables in our regression model, we can see more clearly the relationship between variables and can come to well-reasoned conclusions. We can never know the mind of a public mass shooter and all the twisted reasons why he/she committed the crime. However, from the results of this analysis, we can conclude that firearm availability is a contributing factor enabling the potential public mass shooter. From the point of view of the chances of an event occurring, this example is similar to the tragic case of Sally Clark that we discussed in Chapter 5. The more guns per person that are readily available in a country, the more inevitable it is that sometime, somewhere, someone with a tentative grip on reality is going to obtain one and commit the next public mass shooting. As with the case of Sally Clark, the tragedy of these sort of chance events is that the next victim could be you or someone you love!

9.4 SIMPLE LINEAR REGRESSION

The sample correlation coefficient is a useful statistic for measuring the strength of the linear relationship between two quantitative variables. However, we want to summarize the relationship further by fitting a line to the scatterplot of data points.

By fitting a line to our height-weight data, we are saying that we have determined from the scatterplot that the relationship between height and weight is linear and is consistently linear across all values of height.

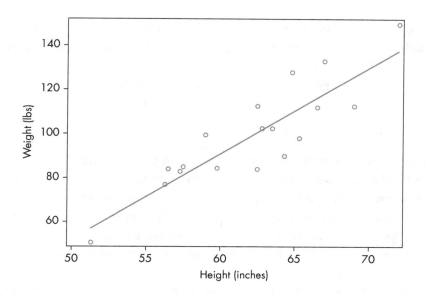

Figure 9.2: Regression Line for Height versus Weight for 19 Children

The statistical technique used to fit a line to our data points is known as **simple linear regression**. The aim of simple linear regression is to choose a line to fit to our scatterplot that gets as close to all the data points as possible. We want to minimize the difference between the actual values of weight, and the predicted values of weight we calculate using the fitted line. For this reason, the line we choose is often called the **line of best fit**. After we fit the line to the data in our example, we will discuss how the line of best fit is chosen.

The regression line we fit to our height-weight data is shown in Figure 9.2. The line runs through the middle of the data, summarizing the relationship between height and weight for the 19 children. The equation for the regression line based on our sample can be written as follows:

$$\hat{y}_i = b_0 + b_1 x_i$$

Predicted Weight = Intercept + Slope x Height
Predicted Weight = −143 + 3.9 x Height

We can use the equation to predict the weight of children of a particular height. For example, for a child who is 62 inches tall, their predicted weight would be calculated as follows:

$$\text{Predicted Weight} = -143 + 3.9 \times (62)$$
$$= -143 + 241.8$$
$$= 98.8 \text{ lbs.}$$

Let's say we selected another child at random from the US population of children. If that child were 62 inches tall, we would predict their weight to be 98.8 lbs. Their actual weight will be lower or greater than this value, but this is the best prediction we can make from our height-weight data. We can think of the predicted weight as our best estimate of the mean weight of children in the population of children who are 62 inches tall. The regression line can be thought of as a line of predicted mean weights across the full range of heights in our sample of 19 children.

The value of 3.9 in the equation for the regression line represents the slope of the regression line known as the sample slope. The sample slope of the line measures by how much the mean predicted value of weight will change for every 1-inch (or unit) change in height. To better understand what this means, let's use the equation of the regression line to predict the weight of a child with a height of 63 inches:

$$\text{Predicted Weight} = -143 + 3.9 \times (63)$$
$$= -143 + 245.7$$
$$= 102.7 \text{ lbs.}$$

The increase in height by one inch from 62 inches to 63 inches has resulted in an increase in the predicted weight from 98.8 lbs. to 102.7 lbs., an increase of 3.9 lbs. Therefore, the sample slope is the increase (or decrease) in the predicted value of the response variable (weight), for every unit (inch) increase (or decrease) in the explanatory variable (height). In other words, for every 1-inch increase (decrease) in height, the mean predicted weight will increase (decrease) by 3.9 lbs.

Before we discuss how the regression line (or line of best fit) is chosen, we should mention what the value of the **intercept**, equal to −143, means. The intercept is the predicted value of weight when height is equal to 0 inches. Of course, no child will have a height of 0 inches or anything close to it. However, the intercept is an important value in determining what the predicted weight will be for any particular height value, and so it should remain as part of the regression equation.

It should also be mentioned at this point that the regression line should not be used to make predictions for weight beyond the range of height values in our sample data. This is called the **danger of extrapolation**. The range of height data values in our example goes from 51.3 inches to 72 inches. We don't have any height measurements beyond this range of values. To use the regression line to make predictions (for weight based on height values) outside this range of values is dangerous. Our predictions would be based on assuming the linear trend we observed exists beyond this range. However, our predictions would be based on no observable data, and in some cases (depending on the measurements taken) could lead to predictions that are far from accurate.

If the line represents the mean predicted values for the response variable weight, across the range of values for the explanatory variable height, then we can think of the vertical distances of the data points (from the line) as deviations from the mean predicted values. These deviations can be thought of in the same way as deviations of individual values around the mean of a single quantitative variable we discussed in Chapter 4. In linear regression, these deviations are known as **residual deviations** or simply residuals.

For example, there are two children in our sample, Janet and Jeffrey, with a height of 62.5 inches. The individual observed weights for Janet and Jeffrey are 112.5 lbs. and 84 lbs., respectively. Using the regression equation, we can find that the mean predicted weight for a child with a height of 62.5 inches:

$$\text{Predicted Weight} = -143 + 3.9 \times (62.5)$$
$$= -143 + 243.75$$
$$= 100.75 \text{ lbs.}$$

Therefore, the individual residual deviations for Janet and Jeffrey are as follows:

$$e_i = y_i - \hat{y}_i$$

residual deviation = observed weight − predicted weight

Janet: 112.5 lbs.: residual deviation = 112.5 − 100.75 = 11.75 lbs.
Jeffrey: 84 lbs.: residual deviation = 84 − 100.75 = −16.75 lbs.

Janet's weight of 112.5 lbs. is above the mean predicted weight of 100.75 lbs., resulting in a positive residual deviation equal to 11.75 lbs. Jeffrey's weight of 84 lbs. is below the mean predicted weight of 100.75 lbs., resulting in a negative residual deviation equal to −16.75 lbs. The residual deviations for Janet and Jeffrey are illustrated in Figure 9.3.

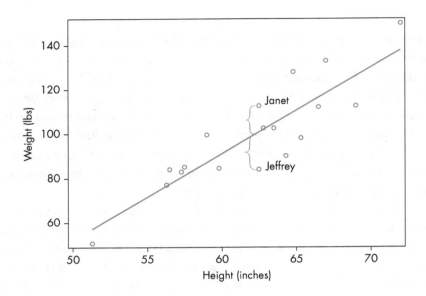

Figure 9.3: Regression Line for Height and Weight–Residuals for Janet and Jeffrey

As was the case with a single quantitative variable, there are both positive and negative deviations for each individual in our sample, representing observed weights above and below the mean predicted weight. The line that we fit to our data is the line that minimizes the sum of the squared residual deviations for all individuals in our sample. This can be written as follows:

$$\min \sum_{i=1}^{n}(y_i - \hat{y}_i)^2$$

Line of Best Fit -> Minimize the Sum of the Squared Residuals

Now that we have some understanding of how we choose our regression line (also known as the line of best fit), we need to think about the effect an individual well above average weight will have on the line fitted to the data. In other words, we need to think about how a data point with a large residual deviation will affect the slope of the regression line.

The underlying mechanics of linear regression are attempting to choose a line that gets as close to each of the data points as possible by minimizing the sum of the squared residuals, the vertical deviations of the data points from the line. A data point far above or below the rest of the data points in the scatterplot will have a large residual deviation and is known as an **outlier**. It will have the effect of pulling the line towards it, with potential to influence the value of the sample slope. In such cases, it is also known as an **influential point**. How much influence the outlier has over the sample slope depends on how large its residual deviation. It also depends on how far it is to the left or right of the sample mean of the explanatory variable.

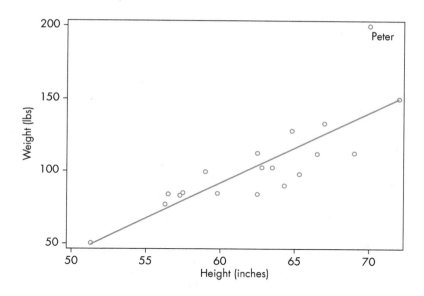

Figure 9.4: Regression Line for Height and Weight with Outlier Peter (70 inches, 200 lbs) Included

To illustrate, let's add another child, named Peter, as the 20th observation to our sample. Peter is 70 inches tall with a well-above average weight of 200 lbs. His data point is considered an outlier. Figure 9.4 illustrates the effect of including Peter in our sample has on the regression line we fit to the data.

The new regression line equation is as follows:

$$\text{Predicted Weight} = -200 + 4.9 \times \text{Height}$$

For a child who is 63 inches tall, the predicted weight will now be as follows:

$$\begin{aligned}\text{Predicted Weight} &= -200 + 4.9 \times 63 \\ &= 108.7 \text{ lbs.}\end{aligned}$$

The new regression line is steeper than the one we previously fit to our height-weight data. The inclusion of Peter in our sample data has the effect of pulling the regression line up toward this data point, increasing the slope of the fitted line from 3.9 lbs. to 4.9 lbs. By including Peter in our sample, the predicted weight for a child who is 63 inches tall has increased from 102.7 lbs. to 108.7 lbs.

The inclusion of Peter as an outlier has resulted in a regression line that does not run through the middle of the data points as well as the regression line obtained without Peter in our sample. The line is still a line of mean predicted values of weight for each value of height. However, as we learned in Chapter 4, individual measurement values far from the mean have

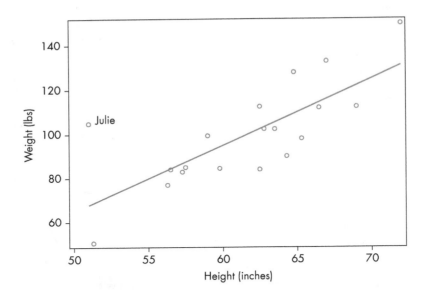

Figure 9.5: Regression Line for Height and Weight with Outlier Julie (52 inches, 105 lbs.) Included

the effect of pulling the mean away from the center of the data, making it a poor measure of central tendency. Our line of mean predicted values is pulled away from the center of the data points resulting in poorer predictions for the response variable weight.

Let's add a shorter-than-average child with a well-above-average weight as the 20th observation in our sample data. Let's say Julie is 52 inches tall and 105 lbs. Her data point is considered an outlier. Figure 9.5 illustrates the effect that including Julie in our sample has on the regression line we fit to our data.

The new regression equation is as follows:

$$\text{Predicted Weight} = -62.3 + 2.6 \times \text{Height}$$

For a child who is 63 inches tall, the predicted weight will be as follows:

$$\text{Predicted Weight} = -62.3 + 2.6 \times 63$$
$$= 101.5 \text{ lbs.}$$

The new regression line is less steep than the one we originally fit to our height-weight data. The inclusion of Julie has the effect of pulling the regression line up toward her data point, decreasing the slope of the fitted line from 3.9 to 2.6 lbs. By including Julie in our sample, the predicted weight for a child who is 63 inches tall has decreased from 102.7 lbs. to 101.5 lbs.

The inclusion of Julie as an outlier has resulted in a regression line that does not run through the middle of the data points as well as the regression line obtained without Julie in our sample data. As a result, the predictions of weight will not be as accurate predictions as they would be without Julie in our sample.

When we added Peter to our sample, his data point (or observation) had the effect of increasing the slope of the regression line from 3.9 to 4.9. When we added Julie, her data point had the effect of decreasing the slope of the regression line from 3.9 to 2.6. The data points for both Peter and Julie are outliers that influence the slope of the regression line.

To fully appreciate how an outlier influences the slope of the regression line, we need to understand that the data point representing the mean height and mean weight (of our height-weight data) is always a point on the regression line. The data point is located in the middle of the line and is known as a **fulcrum point**. The line balances on the fulcrum point like a seesaw, with each of the data points around the fulcrum point playing a role in determining the slope of the line.

When a child of taller-than-average height (and well-above-average weight) was added, the line was pulled up toward that data point from above the fulcrum point, increasing the slope of the regression line. When a child of less-than-average height (and well-above-average weight) was added, the line was pulled up toward that data point from below the fulcrum point, decreasing the slope of the regression line. Let's see what happens to the slope of the regression line when a child of average height (and well-above-average weight) is added to our sample: in other words, when the outlier is directly above the fulcrum point.

The average height of the 19 children is 62.3 inches. Let's add a 20th observation, a child named Brian, 62.3 inches tall and 170 lbs. Figure 9.6 illustrates the effect that including Brian in our sample has on the regression line.

The new regression equation is as follows:

$$\text{Predicted Weight} = -139 + 3.9 \times \text{Height}$$

For a child who is 63 inches tall, the predicted weight will be as follows:

$$\begin{aligned}\text{Predicted Weight} &= -139 + 3.9 \times 63 \\ &= 106.7 \text{ lbs.}\end{aligned}$$

As we can see, the inclusion of Brian in our sample results in the same sample slope of 3.9 that we obtained with the original sample of 19 children. In other words, the outlier has no influence over the sample slope. The fact that the intercept has increased from −143 to −139 means that the predicted weight for a child who is 63 inches tall has increased from 102.7 lbs. to 106.7 lbs., reflecting the change of 4 lbs. in the value of the intercept.

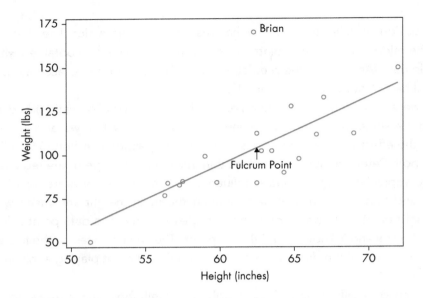

Figure 9.6: Regression Line for Height and Weight with Outlier Brian (62.3 inches, 170 lbs.) Included

The outlier had the effect of pulling the regression line up toward it, but the slope of the line remains unchanged. This is because the outlier was directly above the fulcrum point. How much influence an outlier has on the slope of the line depends on how far it is located to the left or right of the fulcrum point. The further the outlier is to the left or the right of the fulcrum point, the more influence it will have.

It is important to understand the effect that even a single outlying value can have on the resulting regression line that we fit to our scatterplot of sample data. If we have some outlying values in our data, we should first investigate why. They may simply be a measurement error. If not, it is worth investigating further to see why it is an outlier. It could be pointing to a new relationship between the variables that is emerging in the population. In the case of our data on children's heights and weights, it could be pointing to an emerging problem of obesity in children.

Our aim is to estimate the relationship between our two quantitative variables in the population as best as possible. We want to fit a line to our data that best explains the relationship giving accurate mean predictions for our response variable across all values of our explanatory variable. When making statistical decisions about the value of the population slope, one of the necessary conditions is that there are no outliers in our sample that may adversely influence the sample slope. This condition helps to ensure our sample slope is an accurate estimate of the population slope, and the statistical decisions we make regarding the true value of the population parameter are valid.

9.5 CONFIDENCE INTERVALS AND HYPOTHESIS TESTING FOR THE POPULATION SLOPE

As we learned in Chapter 6, in order to make statistical decisions about population parameters using sample statistics, there are necessary assumptions and conditions that our sample data must adhere to. For linear regression analysis, the following are the necessary assumptions and conditions:

Linearity: The assumption that the relationship between the two quantitative variables is linear. We can check this assumption by looking at the scatterplot and deciding whether the trend in the data point should be considered linear.

Independence: The assumption that the residuals are independent of each other. If the data is a representative random sample of individuals from the population, then this assumption should be valid.

Equal Variance: The assumption that there is an even and constant vertical spread of the data points around the regression line, across all values of the explanatory variable. In other words, the data points are equally spread around the regression line we fit to the scatterplot. Once we fit the regression line we can check the scatterplot to see whether this assumption is valid.

Normal Distribution: The condition that the data points must be normally distributed around the regression line for all values of the explanatory variable. Again, once we fit the regression line to the scatterplot, we can check whether this condition is valid by creating a histogram of the residuals. The histogram should be approximately normally distributed.

The above assumptions and conditions are necessary to ensure the validity of our statistical decisions regarding the value of the population slope. If the assumptions and conditions hold, the result of the central limit theorem will be valid. The distribution of possible sample slopes (based on a particular sample size) will be normally distributed around the unknown population slope. Technically, as was the case with a single quantitative variable, the distribution follows a t-distribution and not a normal distribution.

As we have learned previously, the most important calculation necessary for making statistical decisions regarding the value of a population parameter is the standard error. In regression analysis, the standard error measures how far on average we expect our sample slope to deviate from the population slope. The exact calculation of the standard error of the sample slope goes beyond the scope of this book. However, please see the appendix for a brief explanation of the key components involved in the calculation of the standard error of the sample slope.

9.5.1 Confidence Interval

We will now use our sample slope and its corresponding standard error to construct a confidence interval containing a range of plausible values for the population slope. As presented in Chapter 7, the general form of a 95% confidence interval for a population parameter is as follows:

sample statistic ± 2 × (standard error)

Let's return to our 19 children example where we fitted the regression line as follows:

Predicted Weight = –143 + 3.9 × Height

The sample slope is 3.9 lbs. with a standard error equal to 0.5 lbs. (from computer software). The standard error means that we expect a sample slope (based on a random sample of 19 children) to deviate (on average) by 0.5 lbs. from the population slope.

Our 95% confidence interval for the population slope is as follows:

$$b_1 \pm t\,(se\,(b_1))$$

sample slope ± 2 × (standard error)
3.9 ± 2 × (0.5)
[2.9, 4.9]

We can say with 95% confidence that the population slope is somewhere in the interval ranging from 2.9 lbs. to 4.9 lbs. Extending our results to the population, this means that for every 1-inch increase in height, we are 95% confident that the mean increase in weight in the population could be anywhere from 2.9 to 4.9 lbs.

In the college football study we first looked at in Chapter 2, the researcher completed a linear regression analysis looking at the relationship between left hippocampal volume (response variable) and the number of years of football played (explanatory variable). They found a sample slope equal to –43.54 uL. The sample slope means that for every one-year increase in football experience, the predicted hippocampal volume is expected to decrease by –43.54 uL.

The 95% confidence interval for the population slope went from –67.66 to –19.41. We are 95% confident that the population slope could be anywhere from –67.66 uL to –19.41 uL. Extending the results to the population of football players, this means that for every one-year increase in football experience, we are 95% confident that the mean decrease in the value of left hippocampal volume could be anywhere from −67.66 uL to −19.41 uL.

9.5.2 Hypothesis Testing

When we run a hypothesis test (known as a model utility test) for the slope of the regression line, we are testing whether the sample slope provides statistical evidence that the population slope is different from zero. In other words, is there statistical evidence in our data of a linear relationship between our two quantitative variables in the population?

For our 19 children height-weight example, our hypotheses are as follows:

Step I – Hypotheses

Null Hypothesis: population slope is equal to 0 lbs. (H_o: $\beta_1 = 0$)
Alternative Hypothesis: population slope is not equal to 0 lbs. (H_A: $\beta_1 \neq 0$)

By testing this claim, we are saying that we are looking for statistical evidence in our sample data in favor of our alternative hypothesis that the population slope is not equal to zero. A zero slope (horizontal line) would mean that there is no linear relationship between height and weight. No matter what the value of height, the value of weight would always be the same.

Step II – The Model

As long as the necessary assumptions and conditions hold, the central limit theorem states that the distribution of possible sample slopes (based on a particular sample size) follows the normal model, centered at the population slope. Under the null hypothesis, we assume the normal model is centered at 0 lbs. In other words, we are starting with the assumption that there is no linear relationship between height and weight in the population.

The question we are asking is: how likely is our sample slope, given the population slope is 0 lbs.? In other words, how likely is the linear relationship we see between height and weight in our sample, given there is no linear relationship between the two variables in the population?

Using the standard error equal to 0.5, we can calculate the range of possible sample slopes we would expect, given the population slope is equal to 0 lbs.

STUDY FIRST LOOKED AT IN CHAPTER 2:

RELATIONSHIP OF COLLEGIATE FOOTBALL EXPERIENCE AND CONCUSSION WITH HIPPOCAMPAL VOLUME AND COGNITIVE OUTCOMES

Web Link: http://jamanetwork.com/journals/jama/fullarticle/1869211

Search Term: Relationship of Collegiate Football Experience and Concussion

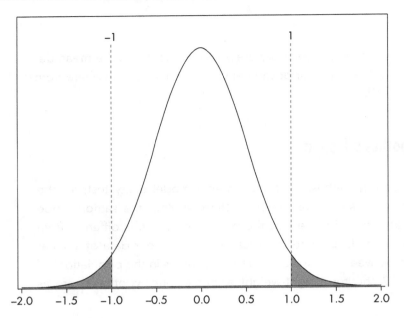

Figure 9.7: Distribution of Possible Sample Slopes

Data source: http://blogs.sas.com/content/sastraining/2014/06/10/producing-normal-density-plots-with-shading/

If the population slope is 0 lbs., we expect 95% of sample slopes (based on sample sizes of 19 children) to fall between −1 lbs. and 1 lbs.

Our sample slope of 3.9 lbs. lies far beyond this range of possible sample slopes, given the null hypothesis is true that the population slope is 0 lbs. It is almost 8 standard errors larger the null value of 0 lbs. Therefore, our sample slope is highly unlikely to occur given that there is no linear relationship between height and weight in the population.

There is strong statistical evidence in our data in favor of the alternative hypothesis that the population slope is not 0 lbs. We formulate our decision making by calculating our p-value.

Step III – Calculations

$$t = \frac{b_1 - \beta_1}{se(b_1)}$$

test statistic = (sample slope − null value)/standard error
= (3.9 − 0)/0.5
= 7.8

The p-value is the probability of getting a sample slope at least as far away from 0 lbs. (as the one we observed), given that the population slope is 0 lbs. In this example, this is equivalent to calculating the probability of getting a test statistic less than or equal to –7.8, or greater than or equal to 7.8 under the normal model. Our p-value turns out to be very small, a value less than 0.0001.

Step IV – Conclusion

With a p-value much smaller than 0.05, we can reject the null hypothesis in favor of the alternative. Based on our sample data, we have found strong statistical evidence of a linear relationship between height and weight in the population of US children.

As we learned in Chapter 8, the smaller the p-value, the stronger the statistical evidence in favor of the alternative. We are simply stating that we have strong statistical evidence that the population slope is not equal to 0 lbs. We are not making a definitive statement regarding what is the true value of the population slope. The 95% confidence interval we calculated, going from 2.9 to 4.9 lbs., gives us a range of plausible values for the population slope.

Let's return to the sample data collected by the 2011–2012 National Health and Nutrition Examination Survey (NHANES). In Chapter 4, we looked at the distribution of the sample of men's height (cm) and men's weight (kg). We will now look at the relationship between these two quantitative variables by creating a scatterplot and fitting a regression line to the data points. The sample had a total of 2,737 men with both the height and weight measurements.

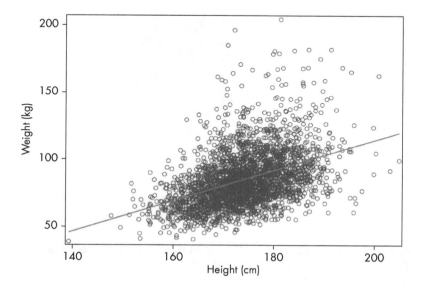

Figure 9.8: 2011–2012 NHANES: Scatterplot of Men's Height and Weight.
https://wwwn.cdc.gov/nchs/nhanes/ContinuousNhanes/Default.aspx?BeginYear=2011.

As we can see from Figure 9.8, there appears to be a linear relationship between height and weight, but it is not a very strong relationship. There is a wide (vertical) spread in the weight values around the regression line. The correlation, as a measure of the strength of the linear relationship, is only 0.42. There is a positive, but weak, linear relationship between height and weight in this data.

The regression equation is as follows:

$$\text{Predicted Weight (kg)} = -110.68 + 1.13 \times \text{Height (cm)}$$

For men who are 180 cm tall, their predicted weight would be calculated as follows:

$$\begin{aligned}\text{Predicted Weight} &= -110.68 + 1.13 \times (180) \\ &= -110.68 + 203.4 \\ &= 92.7 \text{ kg}\end{aligned}$$

Looking at Figure 9.8 again, we can see that there are quite a few men with weights that would be considered outliers. In fact, there are 115 men in this sample weighing 125 kg or greater. We will remove these data points and obtain the resulting regression line based on the new sample size of 2,622 men. Figure 9.9 shows the new scatterplot and regression line with these data points removed.

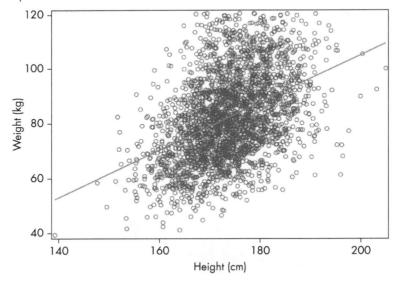

Figure 9.9: 2011–2012 NHANES: Scatterplot of Men's Height and Weight (with Outliers Removed) *https://wwwn.cdc.gov/nchs/nhanes/ContinuousNhanes/Default.aspx?BeginYear=2011.*

The resulting regression equation is as follows:

$$\text{Predicted Weight} = -76.38 + 0.92 \times \text{Height}$$

For men who are 180 cm tall, their predicted weight would be as follows:

$$\begin{aligned}\text{Predicted Weight} &= -76.38 + 0.92 \times (180) \\ &= -76.38 + 165.6 \\ &= 89.22 \text{ kg}\end{aligned}$$

The outliers did influence the slope of the regression line and the resulting predictions for weight quite considerably. By removing the 115 men with weights greater 125 kg, the predicted weight for a man 180 cm tall decreased from 92.7 kg to 89.22 kg. The inclusion of the outliers had influence over the estimated sample slope, increasing the slope from 0.92 kg to 1.13 kg. With the outliers removed, the NHANES data better adheres to the necessary assumptions and conditions for statistical analysis.

The NHANES data is based on a random sample of men from the US population. This should ensure independence in the individual measurement values. Looking at the scatterplot in Figure 9.9, the assumption of a linear trend in the data seems valid. There is fairly even distribution in the vertical spread of the data points around the regression line. The histogram of the residuals shown in Figure 9.10 is approximately normal in shape. The necessary assumptions and conditions for linear regression analysis have been met.

The sample slope is 0.92 kg with a standard error equal to 0.04 kg (from computer software). The standard error means that we expect a sample slope, based on a sample of 2,622 US men, to be (on average) 0.04 kg from the population slope.

Our 95% confidence interval for the population slope is as follows:

$$\boxed{b_1 \pm t\,(se\,(b_1))}$$

$$\begin{aligned}&\text{sample slope} \pm 2 \times (\text{standard error}) \\ &\quad\quad 0.92 \pm 2 \times (0.04) \\ &\quad\quad\quad [0.84, 1.0]\end{aligned}$$

We can say with 95% confidence that the population slope is a value in the interval ranging from 0.84 kg to 1 kg. Extending our results to the population, this means that for every one cm increase in height, we are 95% confident the increase in weight in the population of US men could be anywhere from 0.84 kg to 1 kg.

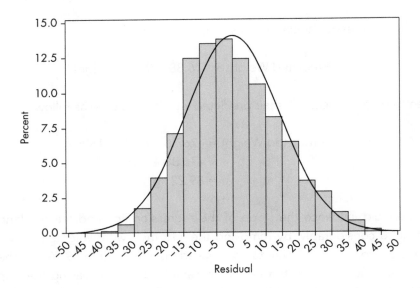

Figure 9.10: 2011–2012 NHANES Men's Height-Weight: Distribution of Residuals. *https://wwwn.cdc.gov/nchs/nhanes/ContinuousNhanes/Default.aspx?BeginYear=2011.*

To test whether or not there is a linear relationship between height and weight in the population, our null and alternative hypotheses are as follows:

Step I – Hypotheses

Null Hypothesis: population slope is equal to 0 kilos (H_0: $\beta_1 = 0$)
Alternative Hypothesis: population slope is not equal to 0 kilos (H_A: $\beta_1 \neq 0$)

By testing this claim, we are saying that we are looking for statistical evidence in our sample data in favor of our alternative hypothesis that the population slope is not equal to 0 kg.

Step II – The Model

Using the standard error of 0.04, we can calculate the range of sample slopes we would expect, given the population slope is equal to 0 kg.

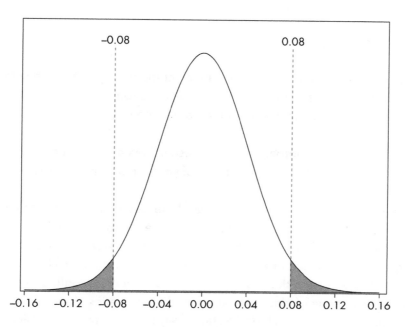

Figure 9.11: Distribution of Possible Sample Slopes

Data source: http://blogs.sas.com/content/sastraining/2014/06/10/producing-normal-density-plots-with-shading/

If the population slope is 0 kg, we expect 95% of sample slopes (based on 2,622 US men) to fall between –0.08 kg and 0.08 kg. Our sample slope of 0.92 kg lies far beyond this range of possible sample slopes, given the null hypothesis is true. There is very strong statistical evidence in our data in favor of the alternative hypothesis that the population slope is not equal to 0 kg. We formulate our decision making by calculating a p-value.

Step III – Calculations

$$t = \frac{b_1 - \beta_1}{se(b_1)}$$

test statistic = (sample slope – null value)/standard error
= (0.92 – 0)/0.04
= 23

The test statistic tells us that the sample slope of 0.92 kg is 23 standard errors greater than the null value of 0 kg. Our p-value turns out to be very small, a value less than 0.0001.

Step IV – Conclusion

With a p-value much smaller than 0.05, we can reject the null hypothesis in favor of the alternative. Based on our sample data, we have found very strong statistical evidence of a linear relationship between height and weight in the population of US men.

Returning to the college football study, the researchers also conducted a hypothesis test looking at the relationship between the number of years of football played and left hippocampal volume.

The sample slope was equal to −43.54 uL, and the resulting p-value was reported as P < .001. The p-value means that the researchers found very strong statistical evidence in their data to reject the null hypothesis of no linear relationship between these two quantitative variables. There is a very small probability of getting a sample slope of −43.54 uL (or one more extreme), given the population slope is equal to zero uL. The 95% confidence interval [−67.66 to −19.41] provides a range of plausible values for the population slope.

In other words, the researchers found a statistically significant (negative) linear relationship between the number of years of football played and left hippocampal volume. They estimated that for every additional year of football played, left hippocampal volume would decrease by 43.54 uL on average. They found with a high level of confidence that the mean decrease in left hippocampal volume could be anywhere between −67.66 uL and −19.41 uL, for every additional year of football played.

9.6 MULTIPLE LINEAR REGRESSION

In our last section, we fit a simple linear regression model to the NHANES men's height and weight data. We used the height variable as one factor for explaining the variability in weights and for making predictions. However, height is not the only factor related to the variability in weight. Other factors related to weight include diet, exercise, and waist size. When we want to include more than one explanatory variable in our linear regression model, the technique is known as multiple linear regression.

Multiple linear regression is a very powerful and versatile technique for conducting statistical analysis. It enables us to build sophisticated models that are more reflective of the complex relationships between factors (or variables) in the world around us. We will never be able to build models that completely reflect the complex relationships between variables in the real world. However, it is possible to build models that enable us to know enough about particular relationships between certain variables in order to make good decisions.

As we saw in the public mass shooting study in Section 9.3, multiple regression enables us to control for and take into account additional factors that may explain (at least some of) the variability in the response variable. We will return to this example once we have a better understanding of the multiple linear regression model.

Returning to the NHANES men's height and weight example, we will add a second explanatory variable, waist size (in cm), to our linear regression model. The resulting linear regression model is as follows:

$$\hat{y}_i = b_0 + b_1 x_{1i} + b_2 x_{2i}$$

Predicted Weight = −138.1 + 0.73 × Height + 0.96 × Waist

We can see from the regression equation that we now have two slope values, one for height equal to 0.73 and one for waist equal to 0.96. These values are also known as the coefficients for height and waist. We can see that the coefficient for height in this model is different from what it was in our simple linear regression, a value equal to 1.13. The effect of including a second explanatory variable in the model has resulted in a different-looking relationship between the variables that we need to reason with and understand. The meaning of the coefficients associated with each explanatory variable has changed in a subtle but important (to understand) way.

In multiple linear regression, the coefficient means that for every unit change in the particular explanatory variable, the mean predicted value of the response variable will change by the value of the coefficient while holding the other explanatory variables in the model constant.

For example, for men who are 180 cm tall with an 80 cm waist, their predicted weight would be calculated as follows:

$$\text{Predicted Weight} = -138.1 + 0.73 \times (180) + 0.97 \times (80)$$
$$= 70.9 \text{ kg}$$

In our simple linear regression model using only height as an explanatory variable, the predicted weight for men 180 cm tall was equal to 92.7 kg. However, this was the predicted weight for all men who are 180 cm tall regardless of their waist size. The predicted weight of 70.9 kg is for men who are 180 cm tall with an 80 cm waist size. Our multiple linear regression model enables us to make weight predictions for men of different heights while holding the value of waist size constant.

For men who are 190 cm tall with an 80 cm waist, their predicted weight would be calculated as follows:

$$\text{Predicted Weight} = -138.1 + 0.73 \times (190) + 0.97 \times (80)$$
$$= 78.2 \text{ kg}$$

The height coefficient equal to 0.73 means that for every 1 cm change in height (for men with a waist size equal to 80 cm), their mean predicted value for weight will change by 0.73 kg. If a multiple linear regression model is suitable (based on checking the assumptions and conditions for linear regression), then this relationship between height and weight will hold for all values of waist size.

Multiple linear regression is a powerful tool for understanding the underlying relationships between variables. By including more than one explanatory variable in our linear model, we can control for certain variables and see the relationship between other variables to the response variable and whether or not those relationships are statistically significant. It also enables us to build sophisticated models for making predictions.

As mentioned previously, with height as the only explanatory variable in our model, for every cm increase in height, the predicted weight would increase by 1.13 kg, the coefficient for height. When we included waist size as a second explanatory variable in our model, the coefficient for height decreased to 0.73. In other words, we see that height does not have as much an effect on our predicted value for weight when waist size is included in the model. In fact, we see from our model that waist size has more effect on the predicted value for weight than height with a coefficient for waist size equal to 0.97. When we think about it, this makes perfect sense. Taller men will generally be heavier than shorter men but this will not always be the case. Some tall men are below average weight (thin and slender), and some shorter men are above average weight (fat and broader). However, for men, a large waist size tends to be a stronger indicator that a person is heavier than the fact that they are tall. Again, good statistical thinking (and common sense) helps us to understand the results of our multiple regression model.

Another statistic that clearly shows the importance of waist in our model is known as the **coefficient of determination**. It measures the proportion (or percentage) of the variability in the response variable that can be explained by a linear relationship with the explanatory variables. It can be calculated by squaring the value of the sample correlation coefficient, symbolized with the letter r. For this reason, it is often referred to as **R-Square**.

In our simple linear regression model with height as our only explanatory variable, R-Square was equal to 0.18. This means that 18% of the variability in individual weight measurements can be explained by a linear relationship with height. In other words, height explains 18% of the variability in weight. When we include waist in our linear model, R-Square increases to 0.86. This means that 86% of the variability in individual weight measurements can be explained by a linear relationship with both height and waist size. It is clear from comparing these values for R-Square that waist size explains a far greater proportion of the variability in weight measurements than height does alone.

Again, this makes perfect sense. For men, there is a stronger positive correlation between waist size and weight than there is between height and weight. In fact, the sample correlation between waist size and weight is equal to 0.86 whereas the sample correlation between height and weight is only equal to 0.42. In other words, waist size is a much stronger and consistent predictor of weight than height, and this fact is reflected in the resulting values of R-Square.

Finally, we will look at the results of our multiple linear regression analysis as it would appear using statistical software. The statistical output is presented in Table 9.2.

Table 9.2: Analysis of Variance (ANOVA)

Source	DF	Sums of Squares	Mean Square	F Value	P-value
Model	2	575192	287596	7557.53	<0.0001
Error (Residuals)	2507	95402	38.05426		
Total	2509	6705594			

Variable Label	DF	Coefficient	Standard Error	T Value	P-value
Intercept	1	−138.1	2.89	−47.8	<0.0001
Height	1	0.73	0.016	44.62	<0.0001
Waist	1	0.97	0.009	109.15	<0.0001

When you look at Table 9.2, you might be surprised to see an analysis of variance (ANOVA) table. In Section 8.8, we discussed how ANOVA is used for comparing several means: to detect whether there is statistical evidence that at least two of the population means we are comparing differ from each other. As a reminder, ANOVA tests to see whether the variability (or difference) between the sample means around what is called the grand mean: the mean of the sample means (known as between group variability) is large relative to the variability of the individual measurements (in each group) around the sample means (known as within group variability).

In a multiple linear regression model, ANOVA is used to test whether the amount of variability in the response variable explained by the (linear) model (with the explanatory variables) is large relative to the amount of variability unexplained by the model—known as the error or residual variability. This test is often called a test of model adequacy. It is a general (or global) test to see if any of explanatory variables in the model explain a statistically significant proportion of the variability in the response variable.

In Section 9.4, we learned that the linear regression model that is chosen (to fit the data) is the one that minimizes the sum of the squared residuals. In Table 9.2, under the heading Sums of Squares, the value of 95402 is the sum of squared residuals that is minimized when fitting our linear regression model. In the context of the ANOVA table, it is referred to as the sum of squares for error, a measure of the amout of variability in height measurements unexplained by the model.

The sum of squares for model (equal to 575192) and for error (equal to 95402) are divided by their associated degrees of freedom (DF), resulting in the mean square for model and error, respectively. The mean square for model (equal to 287596) is divided by the mean square for error (equal to 38.05426), resulting in the F Value (or F Statistic) equal to 7557.53. A more in-depth discussion of why we are making these calculations in the way we do (and what degrees of freedom mean) is beyond the scope of this book. All we need to conclude from our analysis is whether the amount of variability in height measurements explained by the multiple linear regression model is large relative to the amount of variability unexplained by the model. The more variability explained by the model relative to the unexplained variability, the larger the F-statistic will be as a positive number.

The larger the F-statistic, the smaller the p-value. In other words, if the proportion of variability in the response variable explained by the model is large relative to proportion of variability left unexplained in the residuals, the more statistically significant our model will be. If the p-value is less than 0.05, this test of model adequacy indicates that at least one of the explanatory variables in the model is statistically significant. As we can see from the table, our model is highly statistically significant with a p-value less than 0.0001. In other words, there is very strong statistical evidence in our data of a linear relationship between our response variable, height, and at least one of the explanatory variables: weight and waist size.

To determine whether one or both the explanatory variables are statistically significant in the model, we examine the analysis results presented in the bottom half of Table 9.2. What we see here are the results of individual hypothesis tests for height and waist size like we completed in Section 9.5 for the height variable alone. We are testing to see whether the sample coefficients for height (0.73) and waist size (0.97) provide statistical evidence that the corresponding population coefficients are different from zero. With both p-values less than 0.0001, we have strong evidence that the coefficients for both height and waist size are statistically significantly different from zero. In other words, both height and waist size are useful predictors of weight.

As we can see, multiple linear regression is a very sophisticated statistical technique for building models that help explain the relationships between variables in the real world. It enables us to add additional explanatory variables to our linear model to understand the contribution of each variable in explaining the variability in the response variable. It also enables us to add variables as controls so we can better understand the importance of explanatory variables of real interest while controlling for these variables.

Returning to the public mass shooting study discussed at the end of Section 9.3, the researcher was primarily interested in the relationship between firearm ownership rates and the number of public mass shootings across 171 countries. Using a multiple regression model, the researcher was able to control for population size, sex ratio, and percentage of urbanization.

He included two other explanatory variables in the model—homicide and suicide rates. As already discussed, homicide and suicide rates were not found to be statistically significant predictors in the model, thus, eliminating these factors as possibly explaining (at least some) of the variability in public mass shootings across countries.

At stated previously, we are working with observational data, so we can't make definitive causal conclusions. There are many factors (that result in an individual committing a mass shooting) that the multiple regression model has not accounted for. However, the model is useful for helping us see more clearly the association between firearm ownership and the number of public mass shootings across countries. Guns do not directly cause public mass shootings but, as the study suggested, availability certainly makes it easier for the potential mass killer to commit a public mass shooting no matter what his/her reasons for doing so may be.

9.7 CONCLUSION

In this chapter, we looked at how to summarize and analyze the linear relationship between two quantitative variables. We began by looking at a scatterplot of the two variables to decide whether a linear relationship is appropriate. We summarized the linear relationship with a sample statistic known as the correlation coefficient, a measure of the strength (positive or negative) of the linear relationship. Then we fit a line to the scatterplot of data called the regression line, or line of best fit. We used the equation of the line to make predictions for the response variable across the range of values of the explanatory variable. We assessed the effect of outliers on the angle (or slope) of the line of best fit called the sample slope. We used the sample slope, an estimate of the population slope, to create a confidence interval for the population slope. We tested whether the sample slope was statistical evidence that the population slope was different from zero. In other words, we tested whether the linear relationship we found in our sample data was statistical evidence of a linear relationship between the two quantitative variables in the population. Finally, we learned how conduct a multiple linear regression analysis.

9.8 REAL-WORLD EXERCISES

1. Exercise 2 in Chapter 3 asked you to design a survey asking ten questions, both categorical and quantitative. Select data you collected for two quantitative questions you think are linearly related. Decide which variable (resulting from your questions) will

be the explanatory variable and which will be the response variable, and complete the following:

    a. Construct a scatterplot of the explanatory and response variables and calculate (using software) the correlation coefficient. Discuss the form and direction of the relationship by looking at the scatterplot and the value for the correlation coefficient.

    b. Calculate the equation for the regression line. Add the regression line to the scatterplot and discuss what the slope of the line means in terms of the explanatory and response variables.

    c. Calculate the standard error for the sample slope and construct a 95% confidence interval for the population slope. Discuss what the confidence interval means in terms of the original question and the population of interest.

    d. Conduct a hypothesis test to discover whether the population slope is statistically significantly different from zero. Discuss what the results of the hypothesis test mean in terms of the original question and the population of interest.

Before completing parts c. and d., be sure to check the necessary assumptions and conditions for statistical analysis.

2. Choose a place in your town or on your campus where you know many people will be passing by. Ask the next 20 males or females that pass by their height, waist size, and weight.

    a. Construct a scatterplot of the weight and height data you collected, placing the height variable on the x-axis and the weight variable on the y-axis. Briefly describe the relationship between the two variables presented in the scatterplot.

    b. Using statistical software, obtain the intercept and slope for the simple linear regression. Discuss what the slope of the regression line means in context.

    c. Based on what you know about the data you used and how it was collected, discuss how representative you think the data is of the population you are extending the results of your analysis to. Discuss any errors or biases that you might think are an inherent part of the data collected.

3. Find data of interest to you online. From the dataset of your choosing, select two quantitative variables you think might be related.

    a. Decide which variable is the explanatory variable and which is the response variable. Construct a scatterplot for the two quantitative variables. Briefly describe the relationship between the two variables presented in the scatterplot.

b. Using statistical software, obtain the intercept and slope for the simple linear regression. Discuss what the slope of the regression line means in context.

c. Based on what you know about the data you used and how it was collected, discuss how representative you think the data is of the population you are extending the results of your analysis to. Discuss any errors or biases that you might think are an inherent part of the data collected.

4. Based on the descriptive analysis of height-weight data you completed in Q2, using statistical software, conduct a model utility hypothesis test and obtain a confidence interval for the population slope.

   a. Discuss what the results of the hypothesis test mean in context.
   b. Discuss what the confidence interval means in context.
   c. Based on the assumptions and conditions necessary for regression analysis, discuss the validity of the analysis results. Use statistical software to create a histogram of residuals for checking the assumption of normality in the distribution of the residuals.

5. Based on the descriptive analysis of data you found online for Q3, using statistical software, conduct a model utility hypothesis test and obtain a confidence interval for the population slope.

   a. Discuss what the results of the hypothesis test mean in context.
   b. Discuss what the confidence interval means in context.
   c. Based on the assumptions and conditions necessary for regression analysis, discuss the validity of the analysis results. Use statistical software to create a histogram of residuals for checking the assumption of normality in the distribution of the residuals.

6. Based on the data you collected for Q2, using statistical software, complete a multiple linear regression analysis with height and waist size as the explanatory variables and weight as the response variable.

   a. Discuss what the results presented in the ANOVA table mean in context.
   b. Use the statistical output to write out the multiple linear regression equation. Explain what the coefficients of both explanatory variables mean in context.

c. Discuss what the confidence intervals for the population coefficients mean in context.

d. Based on the assumptions and conditions necessary for regression analysis, discuss the validity of the analysis results. Use statistical software to create a histogram of residuals for checking the assumption of normality in the distribution of the residuals.

IMAGE CREDITS

Fig 9.8: Source: https://wwwn.cdc.gov/nchs/nhanes/ContinuousNhanes/Default.aspx?BeginYear=2011.
Fig 9.9: Source: https://wwwn.cdc.gov/nchs/nhanes/ContinuousNhanes/Default.aspx?BeginYear=2012.
Fig 9.10: Source: https://wwwn.cdc.gov/nchs/nhanes/ContinuousNhanes/Default.aspx?BeginYear=2013.

CHAPTER 10

INTEGRITY IN RESEARCH

There are many hypotheses in science which are wrong. That's perfectly all right: it's the aperture to finding out what's right. Science is a self-correcting process. To be accepted, new ideas must survive the most rigorous standards of evidence and scrutiny.

—Carl Sagan

KEY TERMS

P-hacking: running multiple analyses until you find a statistically significant p-value

Risk: a term used in epidemiology defined as the probability that a disease (or outcome) will occur

Relative Risk: a value that compares one group's risk of developing a disease (or outcome) relative to another group

10.1 INTRODUCTION

In the introduction to this book, we talked about how statistics are estimators of truth. How close statistics are to the truth depends on how much error or bias is contained in the measurements we use to calculate our sample statistics. The aim of a good researcher should be to avoid all types of possible error or bias that could be contained in their sample statistics by making every effort to collect quality data. The researcher also needs to be able to reason with the only type of error that can't be avoided: the error in their sample statistic due to sampling variation.

This is quite a rigorous and challenging process. It involves clearly defining your population and selecting a representative sample from that population. It requires collecting valid, reliable, and unbiased measurements. It requires a knowledge of statistical analysis methods so that you choose the appropriate method for the type of data you collected. It requires the data you collect adheres to the necessary assumptions and conditions required for valid statistical analysis. It requires an understanding of how to reason with the results of your analysis. Finally, it requires the researcher to have integrity. The methods of statistical analysis should be used in the pursuit of truth, whatever that truth may turn out to be, and not misused to find statistical evidence for what we *want* to be true.

In this chapter, we will look at some examples of research and ask the following questions:

- Did the researcher collect quality data for analysis?
- Did the researcher check the necessary assumptions and conditions to ensure the analysis was valid?
- Did the hypothesis test have sufficient power?
- Did the researcher conduct multiple tests until they found a statistically significant result?
- Did the researcher use a one-sided or two-sided alternative when conducting their hypothesis test?
- Did the researcher properly present the results of the analysis with confidence intervals and exact p-values?

10.2 DATA INTEGRITY

In Chapter 1 of this book, we talked about how important it is that our sample of individual measurements is representative of the population of individual measurements. We learned that a random sample should be representative of the population.

In Chapter 2, we learned that many errors and biases can be an inherent part of a study leading to poor measurements and in turn sample statistics that are not good estimates of the population parameter of interest. The aim of a good researcher should be to get as close to the truth as possible by ensuring these types of error or bias in their measurements are avoided. When the researcher is ready to analyze the data they have collected, they should feel confident that the only error they have to reason with is the error due to sampling variation.

In Chapter 6, we learned about the necessary assumptions and conditions (regarding data quality) to ensure our statistical analysis is valid. One of the most important assumptions is the independence assumption. We learned that a random sample should ensure independent measurements. However, independence also depends on whether or not the individuals in the study are interacting with each other, possibly affecting each other's measurement value.

In the Fast Diet study from Chapter 2, we found that the sample of individuals used was far from representative of the population to which the researchers were extending the results. However, the weight measurements taken could still have been independent measurements.

To adhere to the independence assumption, the researcher in the diet study should have ensured that there was no interaction between the individuals in the study. If some of the individuals in the fast diet study were meeting regularly to encourage each other to stick to their diets, then the independence assumption would not be valid. This sort of interaction between the individuals in the study would mean their individual weight losses would be correlated with each other. As a result, how the standard error, our measure of sampling variation, was calculated would not be valid, resulting in inaccurate confidence intervals and p-values.

Complete independence between measurements may not always be possible, and some correlation (or dependence) between measurements will not have much effect on the accuracy of our analysis results. However, it should be understood by the researcher that if the individual measurements are far from independent, then the statistical methods we learned in this book should not be used to analyze the data. Our analysis is only as good as the quality of the data.

The quality of the data collected for analysis determines, to a great degree, the quality of the research. The researcher collects and analyzes the data. Therefore, the honesty and integrity of the researcher can be important factors in the overall quality of the data and the research. In the *Journal of Clinical Investigation*, in an article titled "Dishonesty in Scientific Research," the researchers, Nina Mazar and Dan Ariely, discuss how researchers may rationalize dishonesty and the implications for scientific research. They point out that the researchers are not necessarily bad people. The way the research system is set up can lead to bad behavior from otherwise good people:

JOURNAL ARTICLE: DISHONESTY IN SCIENTIFIC RESEARCH

Web Link: https://www.jci.org/articles/view/84722

Search Term: Dishonesty in Scientific Research

> For example, the academic research system rewards statistically significant research findings with prestigious publications, grants, and promotions. Statistically nonsignificant research findings, on the other hand, are almost entirely disregarded, despite the fact that we sometimes learn more from them. Consequently, the system sets up a conflict of interest when, after thousands of dollars of research funding and hundreds of hours of work, one faces null effects.

The paper goes on to discuss other conflicts of interest that can lead to bias in research, including the researchers' need to find what they are looking for. The way the research system is set up can make it very tempting for the researcher to keep running tests until he/she finds statistical evidence for their research hypothesis. In the next section, we will explain why this is an example of poor statistical practice by looking at a controversial piece of research that prompted the field of psychology research to question itself.

10.3 REPLICATION CRISIS IN PSYCHOLOGY

Statistical analysis is a powerful way of reasoning with the world around us. A properly selected random sample of sufficient size can provide us with a sample statistic that is a good estimate of the population parameter of interest. The fact that the population size does not matter when determining how close our sample statistic may be to the population parameter means that a well-conducted study could lead to a treatment that helps millions of people. That is very powerful.

In Chapter 8, we learned the steps for making statistical decisions using hypothesis testing. Hypothesis testing has its strengths, but it also has limitations that can be misused, manipulated or simply misunderstood. In some cases, the researcher misuses the tools of hypothesis testing from a lack of proper understanding of the steps of reasoning involved in the process. We will examine these limitations by looking at one of the studies that led to what has become known as the replication crisis in psychology.

In recent years, there have been a number of publicized studies in psychology and the social sciences with questionable results. This led to a sort of self-examination in psychology that is discussed in the *New York Times* article titled "Many Psychology Findings Not as Strong as Claimed, Study Says."

As the article points out, the researchers that worked on reproducing the findings of the 100 studies found no evidence of wrongdoing on the part of the original researchers. However, they did find that the statistical evidence was not as strong as the researchers had originally claimed.

Out of the 100 studies, 62 had findings that were considered statistically significant but turned out not to be in the replication studies. The replication studies had more subjects than the original studies, giving them greater power to accurately detect the true effect size in the population. It turned out that the larger sample sizes resulted in smaller sample effect sizes, many of which were not statistically significantly different from a zero effect size.

The Reproducibility Project shows the power of replication with sufficient sample size to confirm or repudiate the original findings. We are reasoning with sampling variation, the variation in sample effect sizes from sample to sample. A second (and larger) sample is a powerful way of finding out whether the effect size in the first sample was an accurate estimate of the

---

MANY PSYCHOLOGY FINDINGS NOT AS STRONG AS CLAIMED, STUDY SAYS

The past several years have been bruising ones for the credibility of the social sciences. A star social psychologist was caught fabricating data, leading to more than 50 retracted papers. A top journal published a study supporting the existence of ESP that was widely criticized. The journal *Science* pulled a political science paper on the effect of gay canvassers on voters' behavior because of concerns about faked data.

Now, a painstaking years long effort to reproduce 100 studies published in three leading psychology journals has found that more than half of the findings did not hold up when retested. The analysis was done by research psychologists, many of whom volunteered their time to double-check what they considered important work. Their conclusions, reported Thursday in the journal *Science*, have confirmed the worst fears of scientists who have long worried that the field needed a strong correction.

Web Link: http://www.nytimes.com/2015/08/28/science/many-social-science-findings-not-as-strong-as-claimed-study-says.html

Search Term: Many Psychology Findings Not as Strong as Claimed

true (population) effect size or an inaccurate one due to greater sampling variation in smaller samples.

As stated, the researchers found no evidence of wrongdoing in these studies. However, one study on ESP mentioned in the article that prompted the researchers to complete this project is a good example of how hypothesis testing can be manipulated, misused or misunderstood to the researchers' advantage. The results of this controversial study were discussed in the *New York Times* article titled "Journal's Paper on ESP Expected to Prompt Outrage."

As the news article mentions, the study performed nine experiments on over a thousand subjects. The main hypothesis the researcher was trying to show statistical evidence for was the ability of individuals to have foreknowledge of a future event, known as precognition.

The results of the research were presented in the *Journal of Personality and Social Psychology* in a journal article titled "Feeling the Future: Experimental Evidence for Anomalous Retroactive Influences on Cognition and Affect". Dr. Daryl J. Bem found that 8 out of 9 experiments he conducted rejected the null hypothesis of no precognition in favor of the alternative that precognition exists.

The first problem with this study is how the researcher reasoned to a statistical decision using hypothesis testing. The first step in any hypothesis test is to clearly define your null and alternative hypotheses. In this study, the null and alternative were as follows:

Null Hypothesis: No Precognition Exists
Alternative Hypothesis: Precognition Exists

The next step is to collect a sample of participants who are representative of the general population of interest. In this study, the researcher stated the participants were chosen from the general population, but in fact all participants were university students. The researcher should then run an experiment that tests for precognition and collect his data, which is analyzed to confirm or deny the presence of precognition.

JOURNAL'S PAPER ON ESP EXPECTED TO PROMPT OUTRAGE

One of psychology's most respected journals has agreed to publish a paper presenting what its author describes as strong evidence for extrasensory perception, the ability to sense future events.

The decision may delight believers in so-called paranormal events, but it is already mortifying scientists. Advance copies of the paper, to be published this year in *The Journal of Personality and Social Psychology*, have circulated widely among psychological researchers in recent weeks and have generated a mixture of amusement and scorn.

The paper describes nine unusual lab experiments performed over the past decade by its author, Daryl J. Bem, an emeritus professor at Cornell, testing the ability of college students to accurately sense random events, like whether a computer program will flash a photograph on the left or right side of its screen. The studies include more than 1,000 subjects.

Web Link: http://www.nytimes.com/2011/01/06/science/06esp.html

Search Term: Journal's Paper on ESP Expected to Prompt Outrage

FEELING THE FUTURE: EXPERIMENTAL EVIDENCE FOR ANOMALOUS RETROACTIVE INFLUENCES ON COGNITION AND AFFECT

Web Link: https://www.apa.org/pubs/journals/features/psp-a0021524.pdf

Search Term: Feeling the Future

If the researcher follows these steps exactly (and his data adhered to the necessary assumptions and conditions), then he can use the tools of analysis (confidence intervals and hypothesis testing) we have learned to analyze the data collected.

However, Dr. Bem did not conduct his research in this way. Instead, he went on an exploratory (or fishing) expedition looking for evidence of precognition anywhere he could find it. He ran multiple types of specific experiments and hypothesis tests until he found statistical evidence for precognition. He then used the results of those tests as confirmatory evidence for his general hypothesis that precognition exists.

This is an example of poor statistical practice. A researcher should decide on the research hypothesis they want to test and then look for confirming evidence. They should not go on an exploratory expedition and then use what they find to confirm their original theory. If we run enough tests, we will eventually find statistical evidence for our research (alternative) hypothesis by chance alone. The webcomic looking at the relationship between jelly beans and ache (http:xkcd.com/882/) illustrates this point very well.

When the researcher has a general research hypothesis (like precognition exists), and he runs enough specific experiments, he will eventually find statistical evidence for his general hypothesis. This is due to how we make statistical decisions by reasoning with sampling variation. Every time we run a hypothesis test, we are willing to take a 5% chance of rejecting the null hypothesis when it is true, our type I error rate we learned about in Chapter 8. In other words, we expect one out of every twenty studies to be a false positive (state drug is effective when it is not).

This type of misuse of hypothesis testing is known as **p-hacking**; running multiple experiments until we find a statistically significant result. A second independent replication study is a good way to confirm or refute a claim made by a researcher. However, a researcher should not run multiple exploratory experiments until they find a statistically significant result to use as confirmatory evidence for their general hypothesis.

The researcher should have made adjustments to the p-values to allow for the fact that he was running multiple experiments in this way. These adjustments would have increased the size of the p-values for each hypothesis test and resulted in some (if not all) of the experiments failing to be statistically significant.

The second issue with this study is how the researcher reasoned to a statistical decision that precognition exists. For example, in the first of his experiments, 100 students from the psychology department at Cornell University volunteered and were given extra credit or $5 to participate. The participants sat in front of a computer screen with two curtains appearing on the screen. For each of the 36 trials of the experiment, the participant had to guess which of the curtains would reveal an erotic image. From the student's point of view, the experiment was a test of clairvoyance. However, the specific picture was not placed behind one of the curtains until after the guess was made, making it a test of precognition.

With only two curtains to choose from, the researcher was looking for evidence that the participant could make the correct guess statistically significantly more than 50% of the time, pointing to evidence of precognition. The average proportion of times that all 100 participants guessed correctly, 0.531 (or 53.1%), was the sample statistic used in the analysis. This sample statistic was found to be statistically significantly different from 0.50 (or 50%) with a p-value equal to 0.01.

It is important to point out that the sample percentage of correct guesses of 53.1% was based on 3,600 measurements, with each of the 100 students having 36 guesses. This large sample size made it easier to find such a small effect size of 3.1% to be statistically significant. A large sample size (of good-quality data) is always better at helping us detect the true effect (of, say, a drug for lowering cholesterol), even if the true effect is small. However, a large sample can also be used to declare statistical significance regarding a small to nonexistent effect size of no real meaning or importance. We will end this section with a discussion of what statistical significance in this experiment actually means. However, let's first think a little more about the null and alternative used in this experiment. The hypotheses can be written as follows:

Null Hypothesis: Percentage of erotic images guessed correctly equal to 50%
Alternative Hypothesis: Percentage of erotic images guessed correctly is greater than 50%

One question is why does the researcher choose a one-sided alternative and therefore only look for evidence for what he wants to be true—that precognition exists? If the researcher has the confidence to question how the world works in such a fundamental way, why not be open to the possibility of a phenomenon that is the opposite of precognition? In other words, why not use a two-sided alternative and be open to the possibility that the true percentage of correct guesses is less than 50%?

As we learned in Chapter 8, it is easier to find statistical evidence in favor of a one-sided alternative than for a two-sided alternative; the former having a p-value half the size of the latter.

In eight of the nine experiments that the researcher conducted, the resulting p-values were less than 0.05. In three of these experiments, the p-values would have been greater than 0.05 if the researcher had used a two-sided alternative. If the researcher had also adjusted the p-values for multiple comparisons, how many of the remaining five experiments would have resulted in p-values greater than 0.05?

The last point we will discuss about this research relates to the weakness of hypothesis testing (for making statistical decisions regarding the true effect) when the true effect is small to nonexistent. In the researcher's first experiment using erotic images, the researcher found that 53.1% of the time the participants guessed correctly, a statistically significant result.

When the researcher rejects the null hypothesis in favor of his alternative, he is able to make the claim that precognition exists. However, all the result is really saying is that the statistical evidence in the data points to the possibility that the true percentage of guesses in the population is greater than 50%, if only by a small percentage. Perhaps that is all Dr. Bem wanted to find, to declare precognition exists.

It should also be pointed out that experiments like the ones Dr. Bem conducted are prone to what are known as systematic errors—inaccuracies in the measurement process. As mentioned, the experiment was based on 3600 measurements from 100 students. It would not be surprising if some of these measurements were inaccurate and that is what led to a statistically significant result in some of the experiments conducted.

This example points to how researchers can misuse hypothesis testing to declare that they have found something important. All the researcher had to do to declare precognition exists was to find statistical evidence that the true percentage of guesses is greater than 50%. In this experiment, he collected enough data to show that 53.1% was far enough away from 50% to reach statistical significance and declare that precognition exists. Since the researcher ran multiple tests (without any adjustments of p-values) and used one-sided alternatives, it is not surprising that he found sample statistics in some of his experiments that were statistically significant.

In their critique of this study titled "Why Psychologists Must Change the Way They Analyze Their Data: The Case of Psi," researchers Eric-Jan Wagenmakers, Ruud Wetzels, Denny Borsboom, and Han L. J. van der Maas, based at the University of Amsterdam, allowed for the possibility of such small departures from 50% when analyzing the results of this research. By using alternative statistical analysis tools (beyond the scope of this book) that are more sensitive to such small effect sizes, the researchers found little to no evidence for precognition in eight of the nine experiments conducted. It would not be surprising if the ninth experiment was statistically significant by chance due to sampling variation.

There have been many improvements in how research in psychology is conducted since the replication crisis began. In the FiveThirtyEight website article titled "Psychology's Replication Crisis Has Made the Field Better," the author discusses the changes made to the research process to try and eliminate the type of problems we discussed in the research on extrasensory perception. Journals are asking researchers to preregister their analysis and methods plan in advance as a way of preventing p-hacking. Preregistration is where the journal requires you to specify your analysis plan before you collect your data. This should help reduce the number of

---

WHY PSYCHOLOGISTS MUST CHANGE THE WAY THEY ANALYZE THEIR DATA: THE CASE OF PSI

Web Link: http://citeseerx.ist.psu.edu/viewdoc/download?doi=10.1.1.359.8070&rep=rep1&type=pdf

Search Term: Why Psychologists Must Change the Way They Analyze Their Data

---

ARTICLE: PSYCHOLOGY'S REPLICATION CRISIS HAS MADE THE FIELD BETTER

Web Link: https://fivethirtyeight.com/features/psychologys-replication-crisis-has-made-the-field-better/

Search Term: Psychology's Replication Crisis Has Made the Field Better

false positives found in the field of psychology research. As expected, there has been some resistance to the changes. Forcing a researcher to have integrity in the pursuit of truth is not the same as helping them see why it is important (for themselves and for science) to do so. A researcher with real integrity (and a proper understanding of the tools of statistical analysis) will pursue truth regardless of the system or environment they work under. However, eliminating the temptations in the system to do the wrong thing certainly makes it easier for the good researcher to do the right thing. More researchers are asking the question: "How can we do better?" and that is a good for science and the scientific process.

10.4 THE LOW POWER OF POWER POSING

Another piece of research that received a lot of media attention was to do with what the researcher calls power posing. The idea is very simple: a person can instantly feel more powerful by simply taking on a high-power pose for as little as two minutes. The researcher, Dr. Amy Cuddy, found statistical evidence that a high-power pose raises testosterone levels, lowers cortisol levels, increases the willingness to take risks, and on average makes people feel better. A lot to be gained by taking a couple of minutes to pretend you're Superman or Wonder Woman. The following *New York Times* article discusses the research and the success and notoriety it has brought to Dr. Cuddy.

Amy Cuddy Takes a Stand

The TED conference has made a star of many unlikely people, but perhaps no one more so than Amy Cuddy, a social psychologist and associate professor at Harvard Business School, whose talk promises personal transformation with nary a pill, cleanse or therapy bill.

Her rousing presentation in 2012 at TED Global on what she calls "power poses" is among the most viewed TED Talks of all time (it is No. 2; Sir Ken Robinson's "How Schools Kill Creativity" is No. 1). In its wake, Ms. Cuddy, 42, has attracted lucrative speaking invitations from around the world, a contract from Little, Brown & Co. for a book to be published next year, and an eclectic army of posture-conscious followers.

In her TED Talk, Dr. Cuddy gives a very convincing presentation of her findings. As the news article points out, many people from different walks of life really believe in the power of power posing. Dr. Cuddy stated in

AMY CUDDY TAKES A STAND

Web Link: https://www.nytimes.com/2014/09/21/fashion/amy-cuddy-takes-a-stand-TED-talk.html

Search Term: Amy Cuddy Takes a Stand

> **POWER POSING: BRIEF NONVERBAL DISPLAYS AFFECT NEUROENDOCRINE LEVELS AND RISK TOLERANCE**
>
>
>
> Web Link: http://www.people.hbs.edu/acuddy/in%20press,%20carney,%20cuddy,%20%26%20yap,%20psych%20science.pdf
>
> Search Term: Power Posing: Brief Nonverbal Displays

one talk she gave that "We tested it in the lab—it really works." We will examine the statistical evidence backing up her claims regarding the true effect of power posing. No matter how convincing her claims or the perceived effects that people feel, the strength and validity of the statistical evidence are all that matter.

Dr. Cuddy stated in her TED Talk that she randomized her participants into two groups, though she does not mention the sample size used. According to the journal article titled "Power Posing: Brief Nonverbal Displays Affect Neuroendocrine Levels and Risk Tolerance," there were forty-two participants included in the final analysis. Half the participants were randomized to the high-power pose and half to the low-power pose for two minutes, and then measurements were taken. Dr. Cuddy's results were based on conducting four hypothesis tests:

As hypothesized, high-power poses caused an increase in testosterone compared with low-power poses, which caused a decrease in testosterone, $F(1, 39) = 4.29$, $p < .05$;

Also as hypothesized, high-power poses caused a decrease in cortisol compared with low-power poses, which caused an increase in cortisol, $F(1, 38) = 7.45$, $p < .02$;

Also consistent with predictions, high-power posers were more likely than low-power posers to focus on rewards—86.36% took the gambling risk (only 13.63% were risk averse). In contrast, only 60% of the low-power posers took the risk (and 40% were risk averse), $\chi^2(1, N = 42) = 3.86$, $p < .05$;

Finally, high-power posers reported feeling significantly more "powerful" and "in charge" ($M = 2.57$, $SD = 0.81$) than low-power posers did ($M = 1.83$, $SD = 0.81$), $F(1, 41) = 9.53$, $p < .01$;

As we have already discussed, a researcher should always present a confidence interval for the true effect or population parameter along with the p-value resulting from the hypothesis test. The p-value tells us that there is statistical evidence that the true (population) effect is different from zero or no effect, whereas a confidence interval presents a plausible range of values for the true effect size. There were no confidence intervals presented along with the p-values in the journal article.

Dr. Cuddy found statistically significantly higher testosterone levels, a measure of dominance, in the high-power pose group compared to the low-power pose group. She also found statistically significantly lower cortisol levels, a stress hormone, in the high-power pose group compared to the low-power pose group.

For her primary analysis on risk taking, Dr. Cuddy found that 86.36% of the high-power posers were willing to take a gambling risk compared to only 60% of low-power posers, a difference of 26.36%. She compared and analyzed these percentages using a chi-square test of independence.

The sample percentage difference of 26.36% is quite large and gives the impression that high-power posing does significantly affect people's willingness to take risks. However, we must remember that the variation in sample statistics is large when the sample size is small. In other words, with small sample sizes, we could get a sample effect size that is far from a zero effect size, when the true effect size is in fact zero or no effect. In this research, the sample effect size deviates far enough from zero to be considered statistically significant with a p-value of <0.05.

When a p-value is presented as "p < 0.05", we should question what was the exact p-value. For this analysis of people's willingness to take risks, it turns out that the exact p-value was equal to 0.0495, just below the borderline value for statistical significance of 0.05. When this is the case, we have to question whether or not the researcher went p-hacking until they found the result they desired. In other words, did the researcher run multiple experiments until she found results that provided statistical evidence in favor of her research hypothesis.

Dr. Dana Carney, a coauthor in the study who has since distanced herself from the research, said that some p-hacking did occur when analyzing the data. The use of the abbreviation "DV" is assumed here to mean "dependent variable".

My Position on "Power Poses"

Initially, the primary DV of interest was risk-taking. We ran subjects in chunks and checked the effect along the way. It was something like 25 subjects run, then 10, then 7, then 5. Back then this did not seem like p-hacking. It seemed like saving money (assuming your effect size was big enough and p-value was the only issue)

The self-report DV was p-hacked in that many different power questions were asked and those chosen were the ones that "worked."

Hypothesis testing is like a game of chance. If you keep rolling the dice, it will eventually come up with a winner. The use of a small sample size and p-hacking enabled Dr. Cuddy to find an impressive looking sample effect size (26.23% difference in participant's willingness to take risks), that was found to be statistically significantly different from a zero effect size.

Though the media is slow to respond to debunking of sensational science, thankfully the science community is fighting back against this sort of misuse of the statistical method. In a much less reported replication study titled "*Assessing the Robustness of Power Posing*", the researchers

MY POSITION ON "POWER POSES"

Web Link: http://faculty.haas.berkeley.edu/dana_carney/pdf_My%20position%20on%20power%20poses.pdf

Search Term: My Position on "Power Poses"

ASSESSING THE ROBUSTNESS OF POWER POSING

Web Link: http://journals.sagepub.com/doi/abs/10.1177/0956797614553946

Search Term: Assessing the Robustness of Power Posing

Eva Ranehill, Anna Dreber, and Magnus Johannesson used a much larger sample size (and no p-hacking). They found no statistical evidence for the main hypotheses put forth by Dr. Cuddy.

The researchers followed a similar methodology (in collecting the sample data) as that employed by Dr. Cuddy. One key difference was that the researcher had the participants pose for three minutes instead of two minutes. This difference was not found to have a significant effect on the results. Second, the instructions on how to pose were given via computer, eliminating any potential experimenter error influencing the measurements taken. In Dr. Cuddy's research, the experimenter configured the participants into a high power or low power pose.

Table 10.1: Results from Replication Study: Assessing the Robustness of Power Posing

Measurement	Sample Statistic	Confidence Interval	P-value
Risk-Taking	Mean difference	[−0.085, 0.019]	0.215
Testosterone	Mean difference	[−9.801, 1.647]	0.162
Cortisol	Mean difference	[−0.078, 0.022]	0.272

The researchers used a sample size of 200 participants and found no statistically significant effect size for the three main outcomes of the original research, as shown in Table 10.1. They did find statistical evidence that the participants in the high-power pose felt better on average than participants in the low-power pose. However, this is to be expected. Whether or not these feelings last depends on many other factors besides whether or not you pretend you're Superman or Wonder Woman for a few minutes every day. In her TED Talk, Dr. Cuddy stated that power posing "can significantly change the outcomes of your life." If only life were that easy!

An excellent discussion comparing the effect sizes for risk taking found in the original and replication studies can be found at *Slate* magazine in an article titled "The Power of Power Posing." As the two well-respected statisticians point out, there is a "yawning gap between the news media, science celebrities, and publicists on one side, and the general scientific community on the other. To one group, power posing is a scientifically established fact and an inspiring story to boot. To the other, it's just one more amusing example of scientific overreach."

When it comes to power posing, the yawning gap does not necessarily do much harm. It can make people feel better about themselves (at least for a short while), and some may even fake it until they make it, or until they become it as Dr. Cuddy likes to say. However, there is a more serious

---

THE POWER OF POWER POSING

Web Link: http://www.slate.com/articles/health_and_science/science/2016/01/amy_cuddy_s_power_pose_research_is_the_latest_example_of_scientific_overreach.html

Search Term: The Power of Power Posing

need to prevent researchers from misusing the statistical method for their own personal gain. It undermines science and makes it harder for other researchers to stay focused on the more difficult task of pursuing truth in science. It makes it very tempting for a researcher to simply look for a "truth" that gets them attention from the media and the general public, resulting in recognition and financial reward.

In the next section, we will look at a more serious example where a drug was brought to market that caused heart attacks in a small but significant proportion of the population. The question we will ask is whether or not the drug company knew of the dangers of the drug before its approval for the larger population. We will question the integrity of the individuals involved in making sure the drug got to market despite its apparent dangers.

10.5 VIOXX, HEART ATTACKS, AND THE OPIOID CRISIS

On May 20, 1999, the US Food and Drug Administration (FDA) approved a drug for acute pain, named Rofecoxib, marketed as Vioxx. On September 30, 2004, Vioxx was taken off the market because of increased risk of heart attack for individuals taking the drug. Over the four years that Vioxx was on the market, annual sales revenue was around $2.5 billion. The results of the suits against Merck, the makers of Vioxx, are discussed in the following *New York Times* article.

Merck Agrees to Settle Vioxx Suits for $4.85 Billion

> *Three years after withdrawing its pain medication Vioxx from the market, Merck has agreed to pay $4.85 billion to settle 27,000 lawsuits by people who claim they or their family members suffered injury or died after taking the drug, according to two lawyers with direct knowledge of the matter.*

As the article points out, before the lawsuit settlement in 2007 scientists were debating whether Merck knew about the dangers of the drug before its approval in 1999. The debate centered around the results of a clinical trial published in 2000 that compared Vioxx to the painkiller Naproxen, marketed as Aleve. The debate over the results of the clinical trial is discussed in the following *New York Times* article:

MERCK AGREES TO SETTLE VIOXX SUITS FOR $4.85 BILLION

Web Link: http://www.nytimes.com/2007/11/09/business/09merck.html

Search Term: Merck Agrees to Settle Vioxx Suits

SCIENTISTS AGAIN DEFEND STUDY ON VIOXX

Web Link: http://www.nytimes.com/2006/02/23/business/scientists-again-defend-study-on-vioxx.html

Search Term: Scientists Again Defend Study on Vioxx

Scientists Again Defend Study on Vioxx

With a crucial personal-injury trial over Vioxx set to begin in New Jersey next week, the debate heated up again yesterday about whether Merck understated the drug's risks in a journal article in November 2000.

In a letter published online by The New England Journal of Medicine, 11 scientists who were co-authors of the article said they stood by its original conclusions, despite heavy criticism from the editors of the journal.

....

The trial confirmed that Vioxx seemed to be safer on the stomach, but it also showed that more patients taking Vioxx than naproxen died and that many more suffered heart attacks. As published, the article reported that 17 patients taking Vioxx and 4 taking naproxen had heart attacks during the trial.

As the article points out, the editors at the *New England Journal of Medicine* (where the original journal article was published) published in February 2006 an "expression of concern" regarding Merck's failure to clearly present the risk of heart attacks from taking Vioxx. They point out that the difference in heart attack risk between Vioxx and Naproxen was too large to be due to chance alone. The question we should ask at this stage is why the journal did not express the same concern before it published the original journal article in November 2000.

We will take a look at the original journal article titled "Comparison of Upper Gastrointestinal Toxicity of Rofecoxib and Naproxen in Patients with Rheumatoid Arthritis" to see how complete a picture Merck presented on the risks of heart attack while taking Vioxx.

As already mentioned, Vioxx was primarily marketed as a drug for acute pain. However, this study was conducted to compare the incidence of upper gastrointestinal events for patients on Rofecoxib (Vioxx) and Naproxen (Aleve). The study was a randomized experiment with a total of 8,076 patients enrolled in the study: 4,047 patients were randomly assigned to receive Vioxx, and 4,029 patients received Aleve. This is considered a very large sample size. The sample statistics would be expected to be very good estimates of the population parameters of interest. The abstract on the first page of the journal article presents a summary of the results of the study:

COMPARISON OF UPPER GASTROINTESTINAL TOXICITY OF ROFECOXIB AND NAPROXEN IN PATIENTS WITH RHEUMATOID ARTHRITIS

Web Link: http://www.nejm.org/doi/pdf/10.1056/NEJM200011233432103

Search Term: Comparison of Upper Gastrointestinal Toxicity of Rofecoxib

Rofecoxib and naproxen had similar efficacy against rheumatoid arthritis. During a median follow-up of 9.0 months, 2.1 confirmed gastrointestinal events per 100 patient-years occurred with rofecoxib, as compared with 4.5 per 100 patient-years with naproxen (relative risk, 0.5; 95 percent confidence interval, 0.3 to 0.6; P = 0.005).

As a reminder, risk is a term used in epidemiology (the study and analysis of health outcomes and diseases in populations) defined as the probability that a disease (or outcome) will occur. The risk of gastrointestinal events on Vioxx was 2.1 per 100 patient-years or 0.021. The risk on Aleve was 4.5 per 100 patient-years or 0.045. Relative Risk is a value that compares one group's risk of developing a disease (or outcome) relative to another group. In this case, the relative risk for gastrointestinal events was 0.5, calculated as follows:

$$\text{Relative Risk} = \text{Risk on Vioxx}/\text{Risk on Aleve}$$
$$= 0.021/0.045$$
$$= 0.5$$

The relative risk of 0.5 means that there was a 50% decreased risk of gastrointestinal events on Vioxx compared to Aleve. In other words, the risk of gastrointestinal events on Vioxx was half the risk it was on Aleve. The p-value of 0.005 means that the researchers found strong statistical evidence of a difference between Vioxx and Aleve (regarding the risk of gastrointestinal events) in the population. The 95% confidence interval was [0.3,0.6]. This means that the researchers were 95% confident that the true relative risk in the population could be anywhere from 0.3 to 0.6.

The analysis of the rates of complicated confirmed events, were presented in a similar and consistent way:

The respective rates of complicated confirmed events (perforation, obstruction, and severe upper gastrointestinal bleeding) were 0.6 per 100 patient-years and 1.4 per 100 patient-years (relative risk, 0.4; 95 percent confidence interval, 0.2 to 0.8; P=0.005).

In this case, the relative risk was 0.4 for complicated confirmed events, calculated as follows:

$$\text{Relative Risk} = \text{Risk on Vioxx}/\text{Risk on Aleve}$$
$$= 0.006/0.014$$
$$= 0.4$$

However, when comparing the incidence of myocardial infarction (heart attack), how the relative risk was calculated was not consistent with how it was calculated for the primary results:

The incidence of myocardial infarction was lower among patients in the naproxen group than among those in the rofecoxib group (0.1 percent vs. 0.4 percent; relative risk, 0.2; 95 percent confidence interval, 0.1 to 0.7); the overall mortality rate and the rate of death from cardiovascular causes were similar in the two groups.

The risk of heart attack on Vioxx was 0.004 or 0.4 percent, representing the 17 patients who got heart attacks while on Vioxx. The risk of heart attack on Aleve was 0.001 or 0.1 percent, representing the 4 patients who got heart attacks while on Aleve.

In this case, the relative risk for heart attacks was presented as 0.2, calculated as follows:

$$\text{Relative Risk} = \text{Risk on Aleve}/\text{Risk on Vioxx}$$
$$= 0.001/0.004$$
$$= 0.25$$

The relative risk was calculated by dividing 0.001 by 0.004, the risk on Aleve over the risk on Vioxx, the reverse of how it was calculated for the primary analysis of gastrointestinal events. Both ways of calculating relative risk are valid. However, the researchers should be consistent in how they calculate relative risk for each outcome or event of interest to avoid any confusion or misunderstanding of the results. It is accepted (and proper) practice to list the primary drug of interest first and then the comparator drug. Why did the language change when it came to describing the incidence of heart attacks? Why did the researchers not provide the rate of death from cardiovascular diseases instead of simply stating it was similar in both groups?

The reported relative risk of 0.2 means that there is an 80% decreased risk of heart attack on Aleve compared to Vioxx. In other words, the risk of heart attack on Aleve was one-fifth the risk of heart attack on Vioxx. As shown in our calculation, the actual relative risk was equal to 0.25 (to two decimal places) but the researchers rounded the value to 0.2! If the relative risk for heart attack was calculated in the same way it was for the primary analysis, it would have been equal to 4, a value that is much easier to understand and explain. It simply means (what the data show very clearly) that there was four times the risk of getting a heart attack on Vioxx (17 patients) compared to Aleve (4 patients). A relative risk of 4 would have been much more alarming to the reader of the paper than 0.2 or 0.25.

The 95% confidence interval for the relative risk goes from 0.1 to 0.7. The confidence interval does not include 1, a value that would indicate that the risk of heart attack on both drugs is the same in the population. Therefore, the confidence interval provides statistical evidence that risk of heart attack is significantly higher for patients on Vioxx compared to patients on Aleve. The corresponding p-value would definitely have been less than 0.05. However, the p-value was not presented in the summary of the results or anywhere else in the journal article. The researchers did give the following explanation for the differences in the rates of heart attacks observed:

The rate of myocardial infarction was significantly lower in the naproxen group than in the rofecoxib group (0.1 percent vs. 0.4 percent). This difference was primarily accounted for by the high rate of myocardial infarction among the 4 percent of the study population with the highest risk of a myocardial infarction, for whom low-dose aspirin is indicated. The difference in the rates of myocardial infarction between the rofecoxib and naproxen groups was not significant among the patients without indications for aspirin therapy as secondary prophylaxis.

The researchers state that the rate of heart attacks was "significantly lower" among patients on Aleve, but they do not state that it was statistically significant lower. They explain that the higher rate of heart attacks among patients on Vioxx was due to the higher risk of heart attacks among "4 percent of the study population". However, the study was a randomized experiment. The random assignment of patients to treatments should have ensured that approximately the same percentage of patients with high risk of heart attack were assigned to each treatment group.

The questions that come to mind are as follows:

- Why cause confusion by not consistently calculating relative risk?
- Why would Merck present the relative risk as 0.2 (instead of 0.25) making Vioxx look worse when to comes to the risk of heart attack than it actually was?
- Why did Merck present the p-values for the primary analysis and not for the analysis of heart attacks?
- Why did peer review at the *New England Journal of Medicine* not question how the results were presented?
- How come the FDA did not catch the problems with how the results were presented?
- Did Merck know the drug was causing heart attacks before the drug was brought to market?

If you search for news on Merck and Vioxx, you will find a consensus in the media from the evidence presented after Vioxx was taken off the market that Merck knew the drug was causing heart attacks before it was brought to market. You will find that an estimated 60,000 people died from taking Vioxx, more than the number of Americans who died in the Vietnam War. When an individual kills another person in cold blood, we are shocked and outraged. The individual will be sent to prison as punishment for his/her crimes. The individuals in this case made a calm and deliberate effort to hide from the FDA that Vioxx was causing heart attacks. That is a very cold thing to do given that your decisions are going to cause the deaths of innocent individuals you don't even know. However, nobody went to prison in this case. In fact, many of the individuals involved went on to have very prestigious careers. The questions we must ask ourselves are: Where was the collective moral outrage? How many people even know that this occurred?

When we take a moment to internalize the case of Vioxx, we start to understand the scope of what the individuals at Merck did. None of the 60,000 people deserved to die in this way. The fact it was left to chance for Vioxx to find its victims means that any one of those

NEWS ARTICLE: SACKLERS DIRECTED EFFORTS TO MISLEAD PUBLIC ABOUT OXYCONTIN, NEW DOCUMENTS INDICATE

Web Link: https://www.nytimes.com/2019/01/15/health/sacklers-purdue-oxycontin-opioids.html

Search Term: Sacklers directed efforts to mislead public about Oxycontin

ARTICLE: THE PROMOTION AND MARKETING OF OXYCONTIN: COMMERCIAL TRIUMPH: PUBLIC HEALTH TRAGEDY

Web Link: https://www.ncbi.nlm.nih.gov/pmc/articles/PMC2622774/

Search Term: The Promotion and Marketing of Oxycontin

people could have been your father, mother, sister, or brother. It could have been you. The fact that there were no consequences for Merck beyond a hit to its bottom line means that this will (and has) happened again. The hit to the bottom line is simply the cost of doing business.

Another pain medication approved by the FDA that has led to even more deaths than Vioxx is called Oxycontin, approved for use in 1996. As was the case with Vioxx, Purdue Pharma, the pharmaceutical company that manufactured the drug, went on a massive marketing campaign to physicians and the public without disclosing the fact that drug was highly addictive. Oxycontin went on to become the most highly abused painkiller in the United States and a major contributor to the opioid crisis. As the *New York Times* article titled "Sacklers Directed Efforts to Mislead Public about Oxycontin, New Documents Indicate" points out, the owners of Purdue Pharma were directly involved in the efforts to mislead physicians and the public about the dangers of the drug. Richard Sackler, son of the founder, advised doctors to prescribe the highest (and most profitable) doses and pushed the blame onto patients who got addicted stating, "They are the culprits and the problem. They are reckless criminals."

In a related journal article titled "The Promotion and Marketing of Oxycontin: Commercial Triumph: Public Health Tragedy," presented in the American Journal of Public Health, the researcher Art Van Zee discusses the massive marketing campaign conducted by Purdue Pharma. Also, when discussing the origins of the FDA approval of Oxycontin, he states the following:

> *The FDA's medical review officer, in evaluating the efficacy of OxyContin in Purdue's 1995 new drug application, concluded that OxyContin had not been shown to have a significant advantage over conventional, immediate-release oxycodone taken 4 times daily other than a reduction in frequency of dosing.*

Why did the FDA approve Oxycontin when it was found to no more effective than the pain medications already on the market? Why did the FDA approve Oxycontin when it was found to be highly addictive?

If you search for the latest news on Purdue Pharma and Oxycontin, you will find that collective moral outrage is gathering against the Sackler family. A news article in *The Guardian* titled

"Massive lawsuit says Sackler family broke laws to profit from opioids" discusses a massive lawsuit representing numerous communities and Native American tribes across the United States. In 2019, John Oliver on Last Week Tonight did a piece on the opioid crisis focusing mainly on Richard Sackler. Since Sackler hides away from the public and does not give interviews, Oliver had four great actors play the role of Richard Sackler, to help the viewer internalize how much of an amoral and despicable human being this man truly is.

The Sackler family has made so much money from pushing its drug that it is most likely that none of its members will go to prison. However, as a society, we have to push for the changes necessary to ensure that there are real consequences for these sort of white collar crimes. These people are playing with our lives in the pursuit of money and power. The only thing we know for certain in this world of ours is that we have one very short life to live. No one deserves to lose everything they are and everything they could be in this way. Their families and loved ones should not have to suffer such a premature and painful loss. Punishment for these sort of crimes should amount to more than a hit to the almighty bottom line. The punishment should fit the crime. If that is not the truth, then I don't know what is!

NEWS ARTICLE: MASSIVE LAWSUIT SAYS SACKLER FAMILY BROKE LAWS TO PROFIT FROM OPIOIDS

Web Link: https://www.theguardian.com/us-news/2019/mar/21/sackler-family-500-cities-counties-and-tribes-sue-oxycontin-maker?CMP=Share_AndroidApp_Add_to_Pocket

Search Term: Massive lawsuit says Sackler family broke laws to profit from opioids

VIDEO: OPIOIDS II: LAST WEEK TONIGHT WITH JOHN OLIVER (HBO)

Web Link: https://www.youtube.com/watch?v=-qCKR6wy94U

Search Term: Opioids II

10.6 CONCLUSION

In this chapter, we discussed the importance of questioning the quality of data collected. Statistics are a measure of truth. How good a measure of truth depends on the quality of the data on which our statistics are based. If our data is of poor quality and does not adhere to the necessary assumptions and conditions for analysis, the results of analysis of the data will be of poor quality.

We learned the importance of questioning the researcher's use of the statistical method. The researcher can run multiple tests until they find the result they want (known as p-hacking); run one-sided alternative hypothesis tests to increase the power of their test to find what they are looking for; and use small sample sizes that often result in a large sample effect size that may be far from the true effect size. If they run enough experiments and tests using a small sample size, they will eventually find a sample effect size that looks impressive and is statistically significant.

We learned the importance of questioning how the statistical results of the research are presented in journal articles. As we learned in Chapter 1 with the MMR-autism study, we can't always assume the journal or the media, no matter how reputable, did a good job at reviewing the research results before publishing. We must be able to read and question the results of the research for ourselves. Finally, we learned that there are companies that care more about their bottom line than the lives of the people using their products.

APPENDIX

GENERAL NOTATION

n: sample size
x_i: individual measurement value
$\hat{y}_i$: predicted value (linear regression—Chapter 9)
Σ: Summation Sign

Statistic	Parameter
$\bar{x}$: sample mean	μ: population mean
s: sample standard deviation	σ: population standard deviation
$\hat{p}$: sample proportion	p: population proportion
$\bar{x}_1 - \bar{x}_2$: sample mean difference	$\mu_1 - \mu_2$: population mean difference
$\hat{p}_1 - \hat{p}_2$: sample proportion difference	$p_1 - p_2$: population proportion difference
r: sample correlation coefficient	ρ: population correlation coefficient
b_0: sample intercept	β_0: population intercept
b_1: sample slope	β_1: population slope

CHAPTER 4

The sample standard deviation is obtained by first calculating the sample variance as follows:

Sample Variance = (Sum of Squared Deviations)/(Sample Size − 1)

In formula notation, the sample variance can be written as follows:

$$s^2 = \frac{\sum_{i=1}^{n}(x_i - \bar{x})^2}{n-1}$$

For the NHANES height data, the sample size was 2,742 men; the sample mean height was 174 cm; and the sample standard deviation was equal to 7.7 cm. The following is a partial view of the sample standard deviation calculation, including the individual height values for the two shortest and the two tallest men.

Sample Variance = $\dfrac{(139.5 - 174)^2 + (147.8 - 174)^2 + \ldots\ldots + (202.7 - 174)^2 + (204.5 - 174)^2}{2{,}741}$

= 59.29 cm²

Sample Standard Deviation = Square root of (59.29) = 7.7 cm

CHAPTER 6

Independence Assumption: Sample Size versus Population Size

When collecting our data, we need to ensure that the individual measurements we collect are independent (do not affect each other). Selecting a random sample should ensure independence.

However, the validity of the independence assumption also depends on how large the sample size is as a proportion of the population size. The following is an example to illustrate this issue.

Let's say there is a population of 100 people, of whom 80 have health insurance. This is an unrealistic population size, simply used for demonstration purposes. The probability that an individual selected at random is insured is 80/100, equal to 0.80. We select 20 people at random, and let's say the first 19 people selected are insured. What is the probability that the 20th person is insured, given we know that the first 19 people are insured?

There are 81 people left in the population, of whom 61 are insured. Therefore, the probability that the next person we select is insured is 61/81, equal to 0.75.

For individual measurements to be independent, the probability that any individual selected is insured should be the same for all individuals in the sample selected. However, when the sample size is large relative to the population size, the probability is affected by the outcome from the individuals already selected. The independence assumption will not be valid. We address this problem by ensuring that the sample size is no more than 10% of the population size.

In this example, let's say the population consists of 100,000 people, of whom 80,000 are insured. The probability that an individual selected at random is insured is 80,000/100,000, equal to 0.80. Now, if we were to select 20 people at random, let's say the first 19 people chosen are insured. What is the probability that the 20th person is insured, given we know that the first 19 people are insured?

There are 99,981 people left in the population, of whom 79,981 are insured. Therefore, the probability that the 20th person we select is insured is 79,981/99,981, which is still (approximately) equal to 0.80. The fact that the sample size is small relative to the population size means that the probability that any individual selected is insured remains practically the same for each individual.

As a rule of thumb, by keeping the sample size to no more that 10% of the population size, we can ensure that independence is maintained between individual measurements in our sample and the result of the central limit theorem will still be valid.

Calculating Standard Error

1. Population Mean

The standard error for the population mean is calculated as follows:

standard error = standard deviation/square root of sample size

In formula notation, the standard error can be written as follows:

$$SE(\bar{x}) = \frac{\sigma}{\sqrt{n}}$$

In theory, we use the population standard deviation in this calculation. In reality, this value is unknown, so we will use the sample standard deviation as an estimate of the population standard deviation. In formula notation, the standard error can be written as follows:

$$SE(\bar{x}) = \frac{s}{\sqrt{n}}$$

2. Population Mean Difference

The standard error of the sample mean difference is calculated as follows:

Let PV1 = Population Variance for Group 1   PV2 = Population Variance for Group 2
standard error = square root of ((PV1/Group 1 sample size) + (PV2/Group 2 sample size))

In formula notation, the standard error can be written as follows:

$$SE(\bar{x}_1 - \bar{x}_2) = \sqrt{\frac{\sigma_1^2}{n_1} + \frac{\sigma_2^2}{n_2}}$$

The standard error is obtained by first calculating the variance of individual measurements for both groups in our sample. The variance is the standard deviation squared.

The variance in each group is divided by the sample size for each group and then added together. The square root of this sum is the standard error of the sample mean difference. In theory, we will use the population variances (for both groups) in this calculation. In reality, these values are unknown, so we use the sample variances (for both groups) as estimates of population variances. In formula notation, the standard error can be written as follows:

$$SE(\bar{x}_1 - \bar{x}_2) = \sqrt{\frac{s_1^2}{n_1} + \frac{s_2^2}{n_2}}$$

The t-distribution

The t-distribution was discovered by William Sealy Gosset while working as the quality control manager in the Guinness Brewery in Dublin, Ireland, in the early 1900s. His job was to ensure that the product the Guinness Brewery made was of the utmost quality. Gosset had to work with very small sample sizes; there is only so much Guinness you can taste test in a single day. While doing so, he found that his error rate—the percentage of suspect samples he sent to the lab for further testing that turned out to be of acceptable quality—was higher than he expected.

Gosset was working with the normal (sampling) distribution of sample means based on sample sizes as small as 3, 4, and 5. He assumed that approximately 5% (acceptable Type I error rate) of sample means would be more than two standard errors from the accepted mean quality level by chance alone (due to sampling variation). In other words, he expected only 5% of the samples he sent to the labs for further testing would be found to be good samples. However, the lab reported back that his error rate was closer to 15% in some cases.

To investigate what was happening, Gosset took on the task of sampling numerous samples of size 3, 4, 5, 6, and so on. He calculated the sample means for each sample and used them to construct (by hand) the resulting sampling distributions. What he found was a unique sampling distribution for each sample size used. The resulting sampling distributions were bell-shaped (like the normal distribution) but were fatter in the tails of the curve. However, as the sample sizes increased, the shape of what became known as the t-distribution approached the shape of the normal distribution. With sample sizes of 60 or greater, the two curves are approximately the same distribution. As a result of the fatter tails, a larger percentage of sample means (than expected under the normal distribution) were more than two standard errors from the accepted mean quality level due to sampling variation alone.

Each unique t-distribution (for each sample size used) became associated with a concept known as the degrees of freedom. When running a hypothesis test using a single sample mean, for example, the degrees of freedom are calculated as the sample size minus one. In other words, when working with a sample size of, say, 10, statistical calculations (p-values and confidence intervals) are based off the t-distribution with 9 degrees of freedom.

Finally, Gosset discovered (and later proved mathematically by Ronald Fisher) that the use of the sample standard deviation (in place of the population standard deviation) when calculating the standard error is what resulted in the t-distribution.

Example 6.1

For the men, we calculated a sample mean height equal to 69 inches, a sample variance equal to 9 squared inches and a sample standard deviation equal to 3 inches. For the women, we calculated a sample mean height equal to 65 inches, a sample variance equal to 6.25 squared inches and a sample standard deviation equal to 2.5 inches.

$$\text{standard error} = \text{square root of } ((9/36) + (6.25/57))$$
$$= \text{square root of } (0.36)$$
$$= 0.60 \text{ inches}$$

3. Population Proportion

With hypothesis testing, we begin with a currently accepted or assumed value for the population proportion in the null hypothesis, called the null value. The null value is used in the calculation of the standard error:

$$\text{standard error} = \text{square root of } ((\text{null value}*(1 - \text{null value}))/\text{sample size})$$

In formula notation, the standard error can be written as follows:

$$SE(\hat{p}) = \sqrt{\frac{p(1-p)}{n}}$$

where p is the null value of the population proportion

When constructing confidence intervals, we make no assumptions about the value of the population proportion. We use the sample proportion as an estimate of the population proportion in the calculation of the standard error:

standard error = square root of ((sample proportion*(1 − sample proportion))/sample size)

In formula notation, the standard error can be written as follows:

$$SE(\hat{p}) = \sqrt{\frac{\hat{p}(1-\hat{p})}{n}}$$

Example 6.3

When the number of coin tosses is 100, the standard error is calculated as follows:

standard error = square root of ((0.50*(1 − 0.50))/100)
= square root of (0.025)
= 0.05

When we increase the number of coin tosses to 400, the standard error is calculated as follows:

= square root of ((0.50*(1 − 0.50))/400)
= square root of (0.000625)
= 0.025

4. Population Proportion Difference

With hypothesis testing (discussed in Chapter 8), we start with the assumption that the population proportions for both groups are the same value. Therefore, we calculate a weighted (or pooled) estimate of this value using the sample proportions from both groups.

We calculated the weighted estimate by adding the total number of successes (for example the total number of people who said yes to a question) in both groups, divided by the total sample size for both groups.

The weighted estimate is used in the calculation of the standard error:

standard error = square root of (((weighted proportion*(1 − weighted proportion))/group sample size) + (weighted proportion *(1 − weighted proportion))/group sample size)))

In formula notation, the standard error can be written as follows:

$$SE(\hat{p}_1 - \hat{p}_2) = \sqrt{\frac{p_w(1-p_w)}{n_1} + \frac{p_w(1-p_w)}{n_2}}$$

where p_w is the weighted sample proportion.

When constructing a confidence interval (discussed in Chapter 7), we will make no assumptions about the values of the population proportions in each group. We use the sample proportions from both groups in the calculation of the standard error:

standard error = square root of (((group A sample proportion*(1 − group A sample proportion))/group sample size) + (group B sample proportion*(1 − group B sample proportion))/group sample size)))

In formula notation, the standard error can be written as follows:

$$SE(\hat{p}_A - \hat{p}_B) = \sqrt{\frac{\hat{p}_A(1-\hat{p}_A)}{n_A} + \frac{\hat{p}_B(1-\hat{p}_B)}{n_B}}$$

Example 6.4

In this example, we will use the sample proportions as estimates of the population proportions in the calculation of the standard error as follows:

standard error = square root of (((0.55*(1 − 0.55))/100) + (0.45*(1 − 0.45))/100)))
 = square root of (0.00495)
 = 0.07

When we increase the sample size to 400 for men and 400 women and obtain the same sample proportions for both groups, the standard error would be as follows:

standard error = square root of (((0.55*(1 − 0.55))/400) + (0.45*(1 − 0.45))/400)))
 = square root of (0.00124)
 = 0.035

CHAPTER 7

7.3.1 Population Mean

Example 7.1

We selected a random sample of 36 men, resulting in a sample mean height equal to 69.5 inches, a sample standard deviation of 3 inches. The standard error is calculated as follows:

standard error = standard deviation/square root of sample size
= 3/square root of 36
= 3/6
= 0.5

We selected a random sample of 144 men, resulting in a sample mean height equal to 69.5 inches, a sample standard deviation of 3 inches. The standard error is calculated as follows:

standard error = standard deviation/square root of sample size
= 3/square root of 144
= 3/12
= 0.25

Example 7.2

In Chapter 4, we looked at a sample of 2,742 men's weights from the 2011–2012 NHANES study. The sample mean weight was 85 kg, with a sample standard deviation of 20.6 kg.

The standard error is calculated as follows:

standard error = standard deviation/square root of sample size
= 20.6/square root of 2,742
= 0.39

The NHANES study also contained a sample of 2,794 women's weights. The sample mean weight was 75 kg, with a sample standard deviation of 20.9 kg.

The standard error is calculated as follows:

standard error = standard deviation/square root of sample size
= 20.9/square root of 2,794
= 0.40

7.3.2 Population Mean Difference

Example 7.5

Table 7.1

Sample Groups	Male	Female
Sample Size	2,742	2,794
Sample Mean	85	75
Sample Standard Deviation	20.6	20.9

In formula notation, the standard error can be written as follows:

$$SE(\bar{x}_1 - \bar{x}_2) = \sqrt{\frac{s_1^2}{n_1} + \frac{s_2^2}{n_2}}$$

The standard error is calculated as follows:

$$SE(\bar{x}_1 - \bar{x}_2) = \sqrt{\frac{20.6^2}{2,742} + \frac{20.9^2}{2,794}}$$

$$= 0.56$$

7.3.3 Population Proportion

Example 7.6

We find that 200 out of the 500 voters plan to vote for the candidate, a sample proportion of 200/500, equal to 0.40.

In formula notation, the standard error can be written as follows:

$$SE(\hat{p}) = \sqrt{\frac{\hat{p}(1-\hat{p})}{n}}$$

The standard error is calculated as follows:

$$SE(\hat{p}) = \sqrt{\frac{0.40(1-0.40)}{500}}$$

$$= 0.02$$

7.3.4 Population Proportion Difference

Political Polarization and Media Habits
Population Proportion Difference: Consistent conservatives (CC) versus Consistent liberals (CL)

In Chapter 6, we learned that the standard error (of the sample proportion difference) is the square root of the sum of the variances (of the sample proportions) for both groups we are comparing. Remember that variance is the standard error squared. Therefore, the standard error can be expressed as follows:

standard error = square root of ((variance of sample proportion for CC + variance of sample proportion for CL))

In formula notation, the standard error can be written as follows:

$$SE(\hat{p}_{CC} - \hat{p}_{CL}) = \sqrt{\frac{\hat{p}_{CC}(1-\hat{p}_{CC})}{n_{CC}} + \frac{\hat{p}_{CL}(1-\hat{p}_{CL})}{n_{CL}}}$$

This is the standard formula for calculating the standard error of the sample proportion difference. However, for the Political Polarization and Media Habits study, the researchers provided the margin of errors for each group adjusted upwards for the fact that the response rate was 61%. We will use the margin of errors provided to calculate the standard error of the sample proportion difference in this example.

In Chapter 7, we discussed that the margin of error is approximately twice the standard error. In other words, the standard error for each group is half the margin of error given for each group.

For the CC group, the sample size was 309, with a margin of error equal to 0.072. Therefore, the standard error (of sample proportion) for the CC group is approximately equal to 0.036, and the variance is the square of this value, which is equal to 0.001296.

For the CL group, the sample size was 644, with a margin of error equal to 5.0% (or 0.05). Therefore, the standard error (of sample proportion) for the CL group is approximately equal to 0.025, and the variance is the square of this value, which is equal to 0.000625.

Therefore, the approximate standard error of the sample proportion difference is calculated as follows:

$$\text{standard error} = \text{square root of } ((0.001296 + 0.000625))$$
$$= 0.044$$

CHAPTER 8

8.5 Population Proportion Difference

Example 8.4

In the null hypothesis, we are starting with the assumption that the proportion of men and women that will vote for the candidate are one and the same. Therefore, we need to calculate what is known as a weighted or pooled estimate for this value, based on the sample estimates for both men and women, and use this value in the calculation of the standard error. The pooled estimate is calculated as follows:

weighted sample proportion = (number of men said Yes + number of women said Yes)/ (sample size men + sample size women)

$$= (220 + 180)/(400 + 400)$$
$$= 400/800$$
$$= 0.50$$

The estimate of the standard error will now be calculated as follows:

$$\text{standard error} = \text{square root of } (((0.50*(1 - 0.50))/400) + (0.50*(1 - 0.50))/400)))$$
$$= \text{square root of } (0.00125)$$
$$= 0.035$$

CHAPTER 9

The following is a brief explanation of the key components involved in the calculation of the standard error of the sample slope.

1. Standard Deviation of Data Points Around the Regression Line

As we saw in Chapter 4, the standard deviation is a measure of the average distance any value is expected to deviate from the mean. For regression analysis, the standard deviation is a measure of the average vertical deviation (the residual deviation) of all the data points from the mean predicted value of the response variable, for any particular value of the explanatory variable. In other words, it measures the average residual deviation of points from the line of mean predicted values.

As was the case for a single quantitative variable, the standard deviation is obtained by first calculating variance. For regression analysis, variance is calculated as follows:

$$s_e^2 = \sum_{i=1}^{n} \frac{(y_i - \hat{y}_i)^2}{n-2}$$

Sample Variance = (Sum of Squared Residual Deviations)/(Sample Size − 2)

where residual deviations are the observed value of the response variable minus the predicted value of the response variable, for each sample data value of the explanatory variable. Dividing by the sample size minus two instead of the sample size is a mathematical correction that ensures we are calculating a sample variance that is an unbiased estimate of the population variance. Finally, we take the square root of the sample variance to obtain the sample standard deviation.

The sample standard deviation is in the numerator of the calculation used to obtain the standard error of the sample slope. This means that the larger the residual deviations of the data points around the regression line, the larger the standard error of the sample slope will be. In other words, the wider the vertical spread of the data points around the regression line, the greater the variation in sample slopes from sample to sample.

2. Standard Deviation of the Explanatory Variable

The standard deviation of the explanatory variable is in the denominator of the calculation used to obtain the standard error of the sample slope. This means that the narrower the horizontal spread of the data points, the larger the standard error will be. The wider the horizontal spread, the smaller the standard error will be.

To understand why, remember that the data point (fulcrum point) representing the mean of the explanatory variable and the response variable is always on the regression line we fit to the data. The data points to the left and right of this value have the effect of pulling the line up or down toward them. The wider the horizontal spread of the data points around the fulcrum point, the less wiggle room the sample slope will have to vary from sample to sample.

In other words, the wider the horizontal spread of the data points around the mean of the explanatory variable, the more consistent the sample slope will be from sample to sample, and the smaller the standard error. The narrower the horizontal spread of the data points around the mean, the less consistent the sample slope will be from sample to sample, and the larger the standard error.

3. Sample Size

As was the case with other sample statistics, the larger the sample size, the smaller the standard error will be, and the more precise our sample estimates of the population slope from sample to sample.

INDEX

Symbols

95% Confidence, 40
(Measure of) Central Tendency, 58
(Simple) Linear Regression Model, 189

A

Alternative (Research) Hypothesis, 22, 137
Anonymous Poll/Survey, 49
Average, 6

B

Baseline, 20
Biased Measurement, 16
Blinded, 25

C

Central Limit Theorem, 90
Cohen's d, 157
Conditional Probability, 80
Confidence Interval, 106, 113
Confidence Level, 115
Confidential Poll/Survey, 49
Confirmation Bias, 77
Confounding Factor (or Variable), 7
Convenience Sample, 7
Cross-Sectional Study, 34

D

Danger of Extrapolation, 199
Deterministic Relationship, 189
Double-Barreled Question, 53

E

Empirical Rule, 60
Explanatory Variable, 189

F

False Positive, 80

Fulcrum Point, 203

H

Hazard Ratio, 20
Histogram, 58
Hypothesis Testing, 14, 92, 135

I

Independent Events, 80
Independent Measurements, 106
Influential Point, 200
Interquartile Range, 68

L

Leading Question, 53
Level of Significance, 25, 144
Linear Relationship, 189
Line of Best Fit, 196

M

Margin of Error, 40
Mean, 60
Median, 67
Meta-Analysis, 16
Multiple Linear Regression, 194

N

Normal Distribution, 58
Null Hypothesis, 22, 107, 136
Null Value, 107, 135

O

Observational Study, 7
One-Sided Alternative, 151
Online Poll/Survey, 43
Outlier, 200
Oversampling, 48

P

P-hacking, 228
Placebo, 9
Placebo Effect, 32
Polling, 39
Population, 4
Population Correlation Coefficient, 190
Population Mean, 90
Population Mean Difference, 98
Population Parameter, 5
Population Proportion, 100
Population Proportion Difference, 103
Population Slope, 188
Power of the Test, 174
Probability, 75
Prospective Study, 16
p-Value, 136
P-value, 25

R

Randomized Experiment, 9
Relative Risk, 171, 237
Representative Sample, 6
Residual Deviations (Residuals), 199
Respondent, 39
Response Rate, 40
Response Variable, 189
Risk, 171, 234
R-Square (Coefficient of Determination), 216

S

Sample, 4
Sample Correlation Coefficient, 188
Sample Effect Size, 9, 25
Sample Mean, 90
Sample Mean Difference, 98
Sample Proportion, 101
Sample Proportion Difference, 103
Sample Slope, 188
Sample Statistic, 5
Sample Variance, 62
Sampling Distribution, 91
Sampling Variation, 10, 93
Scatterplot, 188
Simple Linear Regression, 196
Skewed Distribution, 58
Standard Deviation, 60
Standard Error, 90
Standardized Score (z-score), 65
Standard Normal Distribution, 60
Statistical Decision, 10, 14, 90
Statistically Significant, 9, 24
Stratified Sampling, 48
Survey, 39

T

Telephone Poll/Survey, 44
Test Statistic, 143
True Negative, 82
True Positive, 82
Two-Sided Alternative, 150
Type I Error (Alpha Level), 174
Type II Error (Beta Level), 174

V

Valid Measurement, 16
Variable, 5

W

Weighting, 41

CPSIA information can be obtained
at www.ICGtesting.com
Printed in the USA
LVHW061117200121
676912LV00002B/13